# Climate Change Communication and the Internet

This volume provides a timely, state of the art collection of studies examining climate change communication in the era of digital media. The chapters focus on a broad range of topics covering various aspects of both practice and research in climate change communication, ranging from the use of online platforms, to blogs, to social networking sites.

Climate change communication has increasingly moved into Internet-based forums, and this book provides a comprehensive overview of research into Internet and climate change communication. The studies share valuable methodological insights in this relatively new field of research and shed light on the opportunities and challenges underlying the collection and analysis of online climate change-related data.

This book was originally published as a special issue of *Environmental Communication*.

**Nelya Koteyko** is a Reader in Applied Linguistics at Queen Mary University of London, UK. She is interested in the print and digital media representations of science and medicine, and has published widely on the linguistic and sociological approaches to analysing online data.

**Brigitte Nerlich** is Professor Emerita of Science, Language and Society at the University of Nottingham, UK. She is interested in science communication, health communication and climate change communication. She has published widely on issues related to science, language and culture.

**Iina Hellsten** is Associate Professor in Corporate Communication at the University of Amsterdam, Amsterdam School of Communication Research (ASCoR), the Netherlands. Her research focuses on the dynamics of communication networks, in particular in social media settings.

# Climate Change Communication and the Internet

*Edited by*
**Nelya Koteyko, Brigitte Nerlich and Iina Hellsten**

Routledge
Taylor & Francis Group

LONDON AND NEW YORK

First published 2017 by Routledge

2 Park Square, Milton Park, Abingdon, Oxfordshire OX14 4RN
52 Vanderbilt Avenue, New York, NY 10017

*Routledge is an imprint of the Taylor & Francis Group, an informa business*

First issued in paperback 2018

*British Library Cataloguing in Publication Data*
A catalogue record for this book is available from the British Library

ISBN 13: 978-1-138-22386-8 (hbk)
ISBN 13: 978-0-367-07582-8 (pbk)

Typeset in Minion Pro
by RefineCatch Limited, Bungay, Suffolk

**Publisher's Note**
The publisher accepts responsibility for any inconsistencies that may have
arisen during the conversion of this book from journal articles to book chapters,
namely the possible inclusion of journal terminology.

**Disclaimer**
Every effort has been made to contact copyright holders for their permission to
reprint material in this book. The publishers would be grateful to hear from any
copyright holder who is not here acknowledged and will undertake to rectify
any errors or omissions in future editions of this book.

# Contents

# Citation Information

The chapters in this book were originally published in the journal *Environmental Communication*, volume 9, issue 2 (2015). When citing this material, please use the original page numbering for each article, as follows:

**Editorial**
*Climate Change Communication and the Internet: Challenges and Opportunities for Research*
Nelya Koteyko, Brigitte Nerlich & Iina Hellsten
*Environmental Communication*, volume 9, issue 2 (2015), pp. 149–152

**Chapter 1**
*Why Are People Skeptical about Climate Change? Some Insights from Blog Comments*
Paul Matthews
*Environmental Communication*, volume 9, issue 2 (2015), pp. 153–168

**Chapter 2**
*Structure and Content of the Discourse on Climate Change in the Blogosphere: The Big Picture*
Dag Elgesem, Lubos Steskal & Nicholas Diakopoulos
*Environmental Communication*, volume 9, issue 2 (2015), pp. 169–188

**Chapter 3**
*Examining User Comments for Deliberative Democracy: A Corpus-driven Analysis of the Climate Change Debate Online*
Luke Collins & Brigitte Nerlich
*Environmental Communication*, volume 9, issue 2 (2015), pp. 189–207

**Chapter 4**
*Meeting the Climate Change Challenge (MC³): The Role of the Internet in Climate Change Research Dissemination and Knowledge Mobilization*
Robert Newell & Ann Dale
*Environmental Communication*, volume 9, issue 2 (2015), pp. 208–227

For any permission-related enquiries please visit:
http://www.tandfonline.com/page/help/permissions

# Notes on Contributors

**Luke Collins** is a Research Fellow in the Centre for Research in Applied Linguistics, University of Nottingham, UK.

**Ann Dale** is Professor and Canada Research Chair in Sustainable Community Development in the School of Environment and Sustainability at Royal Roads University in Victoria, BC, Canada.

**Nicholas Diakopoulos** is a Research Fellow in the Department of Information Science and Media Studies, University of Bergen, Norway.

**Ran Duan** is a PhD student in the School of Journalism, College of Communication Arts and Sciences at Michigan State University, USA.

**Guy Edwards** is a Research Fellow in the Center for Environmental Studies at Brown University, Providence, USA.

**Dag Elgesem** is a Professor in the Department of Information Science and Media Studies, University of Bergen, Norway.

**Iina Hellsten** is Associate Professor in Corporate Communication at the University of Amsterdam, Amsterdam School of Communication Research (ASCoR), the Netherlands. Her research focuses on the dynamics of communication networks, in particular in social media settings.

**Merav Katz-Kimchi** is a Lecturer at the School of Sustainability, Interdisciplinary Center, Herzliya, Israel.

**Nelya Koteyko** is a Reader in Applied Linguistics at Queen Mary University of London, UK. She is interested in the print and digital media representations of science and medicine, and has published widely on the linguistic and sociological approaches to analysing online data.

**Idit Manosevitch** is a Lecturer in the School of Communication, Netanya Academic College, Israel.

**Paul Matthews** is Associate Professor and Reader in Applied Mathematics in the School of Mathematical Sciences at the University of Nottingham, UK.

**Brigitte Nerlich** is Professor Emerita of Science, Language and Society at the University of Nottingham, UK. She is interested in science communication, health communication and climate change communication. She has published widely on issues related to science, language and culture.

**Robert Newell** is a Researcher in the School of Environment and Sustainability at Royal Roads University in Victoria, BC, Canada.

**J. Timmons Roberts** is Ittleson Professor of Environmental Studies and Sociology at Brown University, Providence, USA.

**Lubos Steskal** is a Post-doctoral Fellow in the Department of Information Science and Media Studies at the University of Bergen, Norway.

**Bruno Takahashi** is Assistant Professor and Research Director at the Knight Center for Environmental Journalism, College of Communication Arts and Sciences at Michigan State University, USA.

# INTRODUCTION

# Climate Change Communication and the Internet: Challenges and Opportunities for Research

The new communicative landscape shaped by the Internet has had profound implications for communication research on climate change and environment. Just as social media technologies have changed the ways we interact, consume, and create in everyday life, these Internet-based platforms, which share attributes both with interpersonal and mass communication, have also opened up new areas for researching public engagement with climate science. In particular, the emergence of tools that enable searching, aggregating, and analyzing online data allows communication researchers (and in fact anybody who is interested) to examine the dynamics of climate change-related debates with an unprecedented breadth and scale. More than a decade ago, Rogers and Marres (2000) used only websites to map the online climate change debate issue networks. Now blogs, reader comments, social media status updates, and a multitude of other online platforms have significantly expanded the research toolkit of scholars interested in studying changing patterns in interpersonal and institutional communication on climate change and the environment.

At the same time, however, these developments have brought new challenges for the study of content, context, and influence of climate change representations, and for the role of different stakeholders from science, politics, and the economy in these online debates. Multiple web-based channels and platforms often make it difficult to assess how and by whom the online content is accessed, used, and co-produced. Increasingly sophisticated algorithms used by service providers may also begin to shape and guide searches and therefore patterns of communication and public understanding. As climate change is a contested area characterized by competition among industry, scientists, NGOs, and policy makers, each of whom aim to frame the issue according to their particular agendas (Schäfer, 2012), online platforms are likely to bring new angles to strategic communication by these stakeholders. It is therefore of increasing importance to examine how the changing patterns of online communication may have impacted on the engagement practices of governmental and non-governmental organizations.

This special issue was designed to stimulate innovative investigations into the relationship between online technologies and climate change communication theories and practices (Nerlich, Koteyko, & Brown, 2010). The articles in this issue draw conclusions from existing research on climate change communication in a changing media landscape and reflect on the next steps that need to be taken to develop this field in the future. Mixed methods research combining different quantitative and qualitative frameworks is of particular interest to this special edition, and the far-reaching implications of social media have inspired contributions spanning different fields including computer science, media sociology, policy studies, computational and corpus linguistics, as well as journalism and communication studies.

Climate skepticism, in the sense of doubting certain aspects of a generally accepted body of climate scientific research, has been studied by sociologists, psychologists, and communication and media researchers. However, so far only very few studies have begun to investigate climate change skepticism as an online phenomenon. Sharman (2014), for example, carried out a detailed study of the climate skeptical blogosphere, focusing in particular on how blogs are linked in networks. In this special issue, two articles take up the challenge of investigating the issue of climate skepticism online. Matthews' study is a first at trying to uncover reasons given by skeptics themselves for adopting a climate skeptical position. The article examines in detail one particularly interesting blog post and comment thread in which readers were invited to discuss their background and views on climate science, followed by a brief qualitative discussion of two other similar blog threads. The article therefore provides some initial insights into motivations for climate skepticism as discussed by climate skeptics and the role played by blogs in motivating a shift to a climate skeptical stance.

The second article by Elgesem, Steskal, and Diakopoulus presents a large-scale network analysis of the English-language blogosphere on climate change and provides an innovative way to combine the network structure of the blogosphere and the content and topics that are discussed in the blog communities. In addition, the paper moves beyond the polarization of climate change debate and provides a detailed view on the differences between skeptics and accepter communities on the blogs. The study concludes that while climate skeptics form one large blog community, there are also several climate change accepter communities. The different accepter communities focus on different parts of the climate change issue. Surprisingly, and in line with Sharman's research (2014), climate change is discussed in relation to science in both the skeptics and the various accepter communities.

Driven by methodological concerns, Collins and Nerlich discuss how the techniques derived from corpus linguistics can be used to examine the ways in which online debates around climate change may create or deny opportunities for multiple voices and deliberation. The authors show that while some aspects of online interaction may discourage alternative viewpoints, the online platforms for user comments can also facilitate dialog and foster engagement. Combining corpus analysis techniques with closer text analysis, the study reveals that the most prolific contributors actively engaged with other online users in the debate through reasoning and argumentation, offering some evidence of deliberation in online discussion threads on climate change.

The role of Internet-based tools for research dissemination and knowledge mobilization in climate change research has been under-researched so far. Newell and Dale take up this challenge in their article that documents the results of a recent "Meeting the Climate Change Challenge" project in British Columbia, Canada. The project examined climate change innovations, and experimented with several Internet-based tools (online case study library, online real-time e-Dialogues and Live Chats, as well as various social media) in disseminating and enhancing climate change communication across the research project and practitioners. The paper is one of the first papers to provide results on the actual use of Internet-based tools and several social media platforms in climate change communication.

The article by Takahashi, Edwards, Roberts, and Duan explores the potential and the pitfalls of establishing an online platform for climate change, in this case the *Intercambio Climático*. This website was established by an NGO in the hope that it would stimulate discussions about climate change and climate change mitigation in Latin America and to influence decision-makers. Its aim was to become not only a very much needed regional go-to website for climate change information but also a means by which Latin American NGOs could gain a voice in international climate politics. Through an analysis of secondary sources and in-depth interviews the authors examine how the website was perceived by participants and what they saw as its successes and its limitations. Although perceptions were mainly positive, participants also voiced concerns about financial and time constraints that are common to many online platforms.

In the final article, Katz-Kimchi and Manosevitch provide an insightful analysis of Greenpeace International's Unfriend Coal protest campaign against Facebook's energy policy. In line with emerging research (in this volume and elsewhere), the authors show how, by providing a platform for mobilizing public support in campaigns, social media are opening up new opportunities for actively engaging diverse audiences in environmental campaigns. In this case study, the NGO used Facebook's affordances as an additional and complementary medium to traditional news media in order to maintain communication with international supporters and engage users through various e-tactics. Bypassing the traditional route of projecting strong agendas and political brands, this successful online campaign enabled the creation of loose public networks around individual action themes and induced cognitive and affective engagement.

The diverse selection of articles in this special issue provides a good illustration of the multiple theoretical, methodological, and empirical approaches subsumed under the encompassing label of "online climate change communication". While the authors chose to focus on different stakeholders, all the contributions share the commitment to methodological and theoretical advancement and the objective of providing critical enquiry into the avenues, contexts and applicability of this emerging line of research.

As such we believe the articles make a valuable step toward developing important directions for research on the role of the Internet in climate change communication research and practice, and we hope to see a continuous growth of this field in the near future.

## Funding

This work was supported by the UK Economic and Social Research Council [grant number RES-360-25-0068] and The Netherlands Scientific Organization, NWO [grant number ORA-NWO 464-10-077].

## References

Nerlich, B., Koteyko, N., & Brown, B. (2010). Theory and language of climate change communication. *Wiley Interdisciplinary Reviews: Climate Change, 1*(1), 97–110. doi:10.1002/wcc.2

Rogers, R., & Marres, N. (2000). Landscaping climate change: A mapping technique for understanding science and technology debates on the World Wide Web. *Public Understanding of Science, 9*, 141–163. doi:10.1088/0963-6625/9/2/304

Schäfer, M. (2012). Online communication on climate change and climate politics: A literature review. *Wiley Interdisciplinary Reviews: Climate Change, 3*, 527–543.

Sharman, A. (2014). Mapping the climate sceptical blogosphere. *Global Environmental Change, 26*, 159–170. doi:10.1016/j.gloenvcha.2014.03.003

Nelya Koteyko, Brigitte Nerlich & Iina Hellsten

# Why Are People Skeptical about Climate Change? Some Insights from Blog Comments

Paul Matthews

*Surveys of public opinion show that a significant minority of the population are skeptical about climate change, and many suggest that doubt is increasing. The Internet, in particular the blogosphere, provides a vast and relatively untapped resource of data on the thinking of climate skeptics. This paper focuses on one particular example where over 150 climate skeptics provide information on their background, opinion on climate change, and reasons for their skepticism. Although these data cannot be regarded as representative of the general public, it provides a useful insight into the reasoning of those who publicly question climate science on the Web. Points of note include the high level of educational background, the significant numbers who appear to have been converted from a position of climate concern to one of skepticism, and the influence of blogs on both sides of the climate debate.*

## Introduction

The subject of public opinion on climate change, and in particular climate change skepticism, is becoming one of increasing interest in the social sciences (Engels, Hüther, Schäfer, & Held, 2013; Hobson & Niemeyer, 2013; Koteyko, Jaspal, & Nerlich, 2012; Painter & Ashe, 2012; Poortinga, Spence, Whitmarsh, Capstick, & Pidgeon, 2011). A useful summary of the subject is given by Pidgeon (2012), in an article introducing a volume of papers on the risks associated with climate change and public perception of these risks. Pidgeon notes that there has been a decline in public concern about climate change in recent years, and that this is a surprise to the academic community. The aim of this paper is to try to provide some answers to this puzzle, based on comments on climate skeptic blogs. Individuals who contribute to online discussions are not representative of the general public, but they do provide one of the more visible aspects of climate skepticism and so studying them may improve understanding of the phenomenon (Sharman, 2014).

The decline in climate concern, and corresponding increase in climate skepticism, has been observed in many opinion polls in several countries. Brulle, Carmichael, and Jenkins (2012) observed that environmental issues are ranked low among issues of public concern in the USA, and that within this category, global warming was ranked lowest of nine topics in one poll. They constructed an index of climate concern, which after a peak in 2007 fell considerably. A study by Smith and Leiserowitz (2012) found an increase in skepticism among the US public from 2002 to 2010. Also Poortinga et al. (2011) report surveys showing increasing skepticism in Europe and the USA, while Whitmarsh (2011) found a doubling in the proportion of the UK public who think climate change has been exaggerated between surveys in 2003 and 2008. Furthermore, a decline in the use of Internet searches for the terms "global warming" and "climate change" since a peak in 2007 was discussed by Anderegg and Goldsmith (2014).

Several papers have looked at the different levels of climate skepticism in different countries, showing significant variation but not a consistent picture. Regarding media coverage, Painter and Ashe (2012) found more news coverage of climate skepticism in the USA and UK than in other countries such as France and China, while Grundmann and Scott (2012) reported greater visibility of skeptical views in the USA and France compared with the UK and Germany. With regard to public opinion, a telephone survey found low prominence of skepticism in Germany compared with the UK and USA (Engels et al., 2013). A comparison of public opinion in nine countries (Hagen, 2013) found that climate change was generally ranked as a low priority compared with other issues such as employment, crime, and education, and that levels of skepticism are significantly higher in the Netherlands, UK, and USA than in Brazil and Mexico.

The relatively large and increasing numbers of people expressing doubts about climate change has naturally prompted studies of the causes of this phenomenon. An investigation into "What skeptics believe" (Hobson & Niemeyer, 2013) acknowledged the importance of this question and studied it using interviews with volunteers, attempting to categorize skeptics into five groups. The study explored what impact deliberative forum discussions may have with mixed results.

Factors that have been suggested as possible reasons for public skepticism include the recent economic downturn; skeptical articles in the media, politics, and worldviews (Corner, Markowitz, & Pidgeon, 2014); fatigue with repetition of the message; or a run of recent cold winters (Capstick & Pidgeon, 2013; Poortinga et al., 2011). It has been found that levels of education and knowledge about science are not important factors (Kahan et al., 2012; Whitmarsh, 2011). An interesting question with regard to media coverage is whether skeptical news articles increase public skepticism, or whether this media presence merely reflects public opinion (Krosnick & MacInnis, 2010).

Brulle et al. (2012) considered several possible drivers for public concern, concluding that the weather and provision of scientific information were relatively minor factors, while the media and political issues are more significant. A study by Lahsen (2013) interviewed a number of climate skeptics with a physics or meteorology background, noting their concerns regarding climate models, observing an association with age and with conservative values. Both of these studies focus on the USA, where political aspects may be more prevalent than elsewhere. In the UK, the supplementary information of Capstick and Pidgeon (2013) reports no significant correlation between climate skepticism and voting Conservative.

One further factor that may have influenced public opinion is the 2009 "Climategate" incident in which many emails between climate scientists were released on the Internet (Grundmann, 2013; Koteyko et al., 2012; Montford, 2012). This led to some criticisms of climate scientists regarding misleading presentation of data, bias in the process of literature review, and withholding of data. The debate about whether this had a major impact is ongoing; Anderegg and Goldsmith (2014) have suggested that the impact was short-lived.

Some recent research has used surveys and interviews to determine what aspects of climate change people are skeptical about (Hobson & Niemeyer, 2013; Poortinga et al., 2011). However, very little work has been done via opinion polls or interviews asking climate change skeptics about the reasons behind their skepticism. Very recently, Capstick and Pidgeon (2013) have begun the investigation of this question, using discussion sessions with participants, but more work of this type is needed to improve understanding of what motivates climate skepticism.

The theme of this special issue is the opportunities provided by the Internet for research into climate change communication. Climate change is discussed very widely across the Internet, via social media, newspaper articles that allow public comments, and blogs. There is therefore a very substantial database of public opinion on climate change and on climate skepticism already available on the Internet that has not yet been extensively mined by researchers in the field. An example of where this has been done is a study of comments made on the website of a UK newspaper before and after climategate (Koteyko et al., 2012). A detailed study of climate skeptic blogs was conducted by Sharman (2014), looking in particular at how they are linked together and which are the most central in the network. A key finding of her work was that these blogs mainly focus on scientific aspects of the debate rather than economic or political issues.

The present paper tries to address reasons for skepticism by looking in detail at one particularly interesting blog post and comment thread in which readers were invited to discuss their background and views on climate science, followed by a brief qualitative discussion of two other similar blog threads. Although these readers are not representative of the general public, this exercise may give some insight into the motivating factors for climate skepticism, in particular the thinking behind those individuals who provide a visible online presence.

### The "Air Vent" Blog and Its Reader Background Thread

The "Air Vent" blog (http://noconsensus.wordpress.com) was set up in August 2008 by Jeff Condon, an aeronautical engineer based in the USA. The "Air Vent" became one of the more active climate skeptic blogs, running over 30 posts in August 2009, one of which received 233 comments. The blog concentrates on the science of climate change, often considering some highly technical details. A particular focus of the blog was the controversy over warming in Antarctica, which led to a publication with Condon as a coauthor (O'Donnell, Lewis, McIntyre, & Condon, 2011). The blog received considerable publicity in November 2009, when it was one of only four blogs to receive the link to the climategate emails (http://noconsensus.wordpress.com/2009/ 11/13/open-letter/#comment-11917). Shortly after this incident, the identity of the blog owner (who had previously written anonymously as "Jeff Id") became public. Since 2012 activity at the blog has declined. The "Air Vent" blog was not mentioned in the review of climate skeptic blogs carried out by Sharman (2014).

In April 2010, Condon launched a "Reader Background" post (http://noconsen- sus.wordpress.com/2010/04/21/reader-background/), at the suggestion of a reader. The post proposed to readers "a discussion of our various backgrounds and how we came to be interested in climate science." The contents of this thread provide an interesting insight into the background, opinions and attitudes of those who actively participate in the climate skeptic blogosphere, and the responses given form the basis of the following sections. It is important to emphasize, however, that any analysis of such comment threads is unlikely to be able to address the question of the reasons for skepticism in the general public, since contributors to the thread are a small self-selected group. Also, in making a study of blog comments, there is the assumption that the comments are made in good faith and can be taken at face value; there is of course a possibility that some contributors may have, for example, exaggerated their credentials.

At the time of writing (October 2013), the Reader Background thread contains 251 comments. Among these comments, 166 individuals can be identified as responding to the request (the remainder being commentary, responses to comments from others, or spam). Of these 166 responses, eight provide some background information about themselves but do not give a clear view on climate change. A further four comments indicate support for the mainstream view of climate scientists that warming is caused predominantly by emission of greenhouse gases, so cannot be categorized as skeptics. This leaves 154 individuals who express some degree of skepticism regarding climate change. Here, skepticism is interpreted in a very broad sense, incorporating doubts about future warming, the opinion that past warming is mostly natural, criticism of climate science, or the view that warming will be beneficial. The vast majority of these 154 comments were posted in April 2010, but 13 were written later in 2010, 2 in 2011, and 4 in 2012.

## The spectrum of skeptical views

Attempts to categorize the views of climate skeptics are fraught with difficulties, since there are so many different views and attitudes associated with climate skepticism; a continuous and multidimensional spectrum of opinions cannot be neatly divided into a small number of pigeonholes. Hobson and Niemeyer (2013) describe five overlapping groups, based on interviews with members of the public in Australia. One taxonomy, originally proposed by climate scientist Stefan Rahmstorf, distinguishes between "trend" skeptics (who doubt that warming is occurring), "attribution" skeptics (who doubt the link to human activity), and "impact" skeptics (who accept that warming is occurring due to human activity but doubt that it will lead to serious detrimental impacts); this is used by Poortinga et al. (2011), but these authors note that the categories are strongly linked and that the public does not clearly distinguish between them, so this description may be flawed. Two main categories are used by Capstick and Pidgeon (2013): "epistemic" skepticism, questioning the science, and "response" skepticism, questioning the value of taking action on climate change. Focusing more on the views of scientists, Lahsen (2013) distinguishes between "mainstream skeptics," exhibiting moderate levels of skepticism, and "contrarians," who are more categorical and typically are also skeptical of other environmental issues.

When examining the blog comments, it was difficult to distinguish clearly between the attribution/impact categories of skepticism. More apparent was the degree of skepticism. The following three-way categorization of comments on the Reader Background thread was adopted, in an attempt to give an overview of the range of opinions expressed. This system is still problematic, since the boundaries between the groups are not very clearly defined, but seems to be the most suitable classification in this context.

## Lukewarmers

The term "lukewarmer" is quite commonly used in the climate blogosphere but does not appear to be used in the social science literature. The term is used to describe themselves by 17 of the 154 individuals classified as skeptical. A further eight commenters express lukewarm sentiments, so that approximately one in six of the respondents can be described in this way. Typically, lukewarmers accept that human emissions of carbon dioxide have warmed the planet significantly and will continue to do so in the future. However, they believe that the level of warming will be lower than that predicted by many climate scientists, and that the global warming scare has been exaggerated. They also often express the view that such a moderate level of warming will be beneficial rather than damaging.

Based on this description, it could be argued that lukewarmers should not be regarded as skeptics. But it should be noted that they often express very critical views of climate science, for example, by describing the behavior of some scientists as being appalling, or saying that climate models are useless. One individual refers to himself (perhaps ironically) as a "lukewarm denialist," and others describe themselves as skeptics and lukewarmers.

## Moderate skeptics

The majority of skeptics commenting on the Reader Background thread fit into this broad category. This is characterized by views that warming of the climate is not a problem, that a large proportion of past warming is due to natural processes, that the threat posed by climate change has been greatly exaggerated, that much of the science of climate change is of poor quality, and that some climate scientists have behaved unprofessionally. The distinction between this group and the previous category is not always clear from the comments. The main difference is that "lukewarmers" accept that greenhouse gases are a major driving factor behind global warming, and that warming is likely to continue, while the "moderate skeptics" do not. Skepticism may be expressed explicitly, for example, the declaration that "I am a skeptic," or through expression of views such as those listed above, or implicitly, for example, by stating support for a skeptical blog.

## Strong skeptics

The term "strong skeptics" is used here for those who express the opinion that climate activists or climate scientists are in some way dishonest or fraudulent. This is a relatively small group within the skeptics on the Reader Background thread. The word "scam" is used by six people in relation to climate change, "lie" or "lies" appear four times, three individuals apply the word "fraud," and "dishonest" only appears once. In total, only 13 contributors use such terminology, less than 10% of the total number of skeptics.

## Real names or pseudonyms?

An interesting question with regard to comments on climate blogs is whether contributors use their real names or comment anonymously using a pseudonym. One might suspect that use of real names leads to a more civil discussion, and research on newspaper comment sections supports this hypothesis (Santana, 2013).

Of the 154 identified skeptics, 53 use what appears to be a full real name either in the form of a first name and last name or (in four cases) initials and surname. Thus approximately two-thirds of the skeptic contributors write anonymously. One commenter draws attention to the fact that a significant number have used their real name, and another remarks that those who are retired or self-employed may be more comfortable doing so. One gives university employment as a reason for using a pseudonym.

## Education and background

A particularly striking aspect of the comments on the Reader Background thread is the high level of educational background. Forty of the 154 skeptics state that they have a Ph.D. degree. Of these, 12 are in chemistry; 8 in some form of engineering; 8 in physics; 4 in mathematics; 3 in biological sciences; and the remaining 5 in arts subjects, computer science, economics, or unspecified. A further 11 contributors cite an M.Sc. degree as their highest educational level, while 3 have an MBA. Of the remainder, 46 have a B.Sc. and 14 a BA degree. Among those who do not specify any educational details, technical backgrounds predominate, such as engineering, electronics, and computer software. Some comments on the thread refer to the number of academic qualifications, while others comment ironically on their own lack of qualifications. The blog owner expresses the hope that those without formal qualifications will not be intimidated.

## Always skeptical or converted to skepticism?

An interesting aspect of climate change skepticism is the question of whether those who hold skeptical views have always felt this way since they first heard of the subject or have changed their opinion and become skeptical. A significant number of comments on the Reader Background thread give a clear indication on this point, although many do not.

There are 26 comments (17%) indicating a clear predisposition to skepticism regarding climate change. These individuals either simply state that they have always been skeptical, or that they have a natural tendency to be doubtful or contrarian, or that the idea just sounded wrong. In some cases this results from familiarity with previous scares; this point will be discussed further below.

On the other hand, 42 contributors (27%) indicate clearly that they were originally concerned about anthropogenic global warming (or in a few cases agnostic on the issue), but became increasingly skeptical as they investigated the problem more deeply. Some skeptics indicate that they were originally totally convinced by the arguments for climate change, as a result of their own environmentalist views or, for example, as a result of seeing the film "An Inconvenient Truth," before changing their views as a result of one of the factors that will be discussed in the following section.

## Reasons Given for Skepticism

Most of the 154 identifiable skeptics on the Reader Background thread give some reasons for their skepticism. In some cases one factor dominates, and in others there is a wide range of reasons given. Only eight of these individuals did not provide any clear reasons.

Any categorization of these reasons is difficult and unavoidably subjective, but after closely reading the comments the following main categories were identified.

## Hype, exaggeration, and alarmism

The concept that skepticism may be driven in part by the view that claims regarding climate change are exaggerated has been supported by the work of Whitmarsh (2011); this study found that media alarmism and exaggeration are a significant factor (among many others) and an increasingly important one.

A recent paper addressed in some detail the issue of "The ironic impact of activists" (Bashir, Lockwood, Chasteen, Nadolny, & Noyes, 2013) showing with a detailed sequence of studies that activists can be seen as militant or eccentric, in which case their actions can backfire.

In the Reader Background thread the word "hype" is mentioned by five individuals, while nine use the word "exaggerated." The term "alarmism" is used nine times. Overall, 32 contributors to the thread indicate that this may have been a factor pushing them toward a skeptical view on climate change. The discussion of alarmism is generally directed toward the media with its sensationalist headlines or activists engaging in scaremongering, although some individuals accuse climate scientists of being alarmist, for example, by linking individual disasters such as hurricanes to global warming.

## Previous scares

Somewhat related to the preceding issue of exaggeration is another factor that may lead to climate change skepticism: personal experience of earlier predictions of disaster that have turned out to be either groundless or exaggerated. This issue is likely to be more relevant to older individuals and therefore may be a factor in the observed correlation between age and skepticism (Poortinga et al., 2011; Whitmarsh, 2011).

Fifteen of the 154 identified skeptics mention this issue as an influence on their views. In some cases there are general comments regarding decades of previous catastrophic predictions, while others mention specific events. Foremost among these is the ice age scare or global cooling scare of the 1970s, mentioned explicitly by six contributors. The population explosion scare associated with the work of Paul Ehrlich is mentioned twice.

The possible link between earlier environmental scares and climate skepticism does not seem to have been discussed in the social science literature, although Capstick and Pidgeon (2013) quote one skeptic raising this issue.

## Politics

The link between political opinions and concern about climate change has been well established (Hmielowski, Feldman, Myers, Leiserowitz, & Maibach, 2013; Krosnick & MacInnis, 2010; Poortinga et al., 2011), with those holding more conservative views less likely to be concerned about climate change than those with left-wing views.

However, a study by Engels et al. (2013) shows a negative correlation between political participation and skepticism.

On the Reader Background thread, politics is mentioned regularly, but it can be difficult to distinguish between those who comment on political aspects and those for whom this was a significant factor in forming their skeptical views. Approximately 34 of the 154 skeptics appear to be in the latter category. Typically, these individuals complain that climate science has become politicized, or that some of those promoting action on climate change have a political agenda, or that some people are using concern over climate change as a means of spreading left-wing ideology, increasing centralized control, or increasing taxes. However, 10 of these 34 individuals refer to their own political views, implying either that their own views do not fit well with actions to address climate change, or that they like the political opinions expressed in skeptical blogs.

In the social science literature cited above, the political aspect of climate change views is generally portrayed as a left/right or liberal/conservative split. However, the terminology used in comments from climate skeptics on the Reader Background thread suggests that this may not be the most important aspect. The term "left" (in the political sense) only appears in four comments, three noting the association between left-wing views and climate change, and one referring to his own views. The word "right" only appears twice, and "conservative" four times. But the term "libertarian" is used on nine occasions, with seven individuals explicitly linking themselves with a libertarian political viewpoint. So libertarianism, with its emphasis on individual freedom and opposition to state control, may provide a better fit with climate skepticism than the conventional left/right view of politics. Note that libertarian views are not necessarily conservative (see, for example, http://www.politicalcompass.org/). Of course, counting words such as "left" and "right" is a crude indicator of political viewpoints, as commenters may be unlikely to use these words explicitly. It may be possible to infer political views implicitly from the tone of the responses, but this is problematic since such inference is subjective.

## Climategate

The climategate incident, as noted earlier, was publicized via the "Air Vent" blog in November 2009, five months before the posting of the Reader Background thread. It is thus to be expected that the incident would be commented on extensively; in fact 30 of the 154 skeptical comments refer to climategate, perhaps fewer than one might have expected. Of these, 26 use the explicit term "climategate," with the remaining four referring to "Climatic Research Unit emails."

On reading these 30 comments carefully it appears that the climategate incident was a significant factor in creating skepticism for only 13 individuals. These people express some shock at the revelations in the emails or say that it encouraged them to look into climate science in more detail.

For the remaining 17 commenters, the climategate issue is mentioned as an afterthought toward the end of the comment, after the main reasons for skepticism have already been given. Some individuals state that they were not surprised by climategate, or that it reinforced their existing views. Hence it seems that for most of the climate skeptics commenting on this thread, climategate was not a major factor in forming their skeptical views, since these views had been formed prior to the incident.

*Poor science?*

One of the main reasons given for skepticism is the opinion that some aspects of climate science are of poor quality, or unjustified or insufficiently rigorous. Approximately 60 of the 154 skeptics give this as a factor motivating their skepticism. In some cases this is simply a general statement, for example, saying that they looked into the science and did not find it convincing, but most commenters provide more specific criticisms.

Foremost among these science issues, mentioned by 30 of the climate skeptics, is the "Hockey stick" picture, which shows apparently unprecedented warming in recent decades. This arises from reconstructions of past temperature from proxy data and has been much criticized by climate skeptics, being the earliest issue on which skeptical bloggers achieved scientific publications (McIntyre & McKitrick, 2005). The hockey stick has become an iconic image for the scientific side of the climate debate; two books have been written on this controversy, from opposite perspectives (Mann, 2012; Montford, 2010). Many other aspects of the science are criticized, including overreliance on models, quality of data and data handling, statistical methodology, and computer coding.

This emphasis on scientific details is to be expected given the scientific focus of this particular blog and the high level of scientific and technical background of the commenters, noted in the previous section. In this regard, the blog is typical: a survey of climate skeptic blogs found that the most prominent of these generally tend to concentrate on scientific aspects of the climate debate (Sharman, 2014).

*Influence of blogs*

Other blogs are mentioned frequently by climate skeptics on the Reader Background thread: 85 of the 154 skeptics refer to other blogs, in most cases naming specific blogs but in a few cases as a general remark. Unfortunately, it is difficult, if not impossible, to distinguish between those for whom reading skeptical blogs was a major factor in forming their skepticism, those whose doubts about climate change were strength-ened by skeptical blogs, and those who turned to the skeptical blogs after forming their own skeptical opinions.

The number of citations of the different blogs is shown in Figure 1. The most frequently cited blog is "Climate Audit", mentioned 57 times, indicating its premier status among blogs questioning climate science. Among other skeptical blogs, the "Air Vent" blog hosting the thread is mentioned 35 times and the "Watts Up With That?" blog 34 times. "John Daly," of interest since he was probably the first climate skeptic to set up a website, is referred to four times.

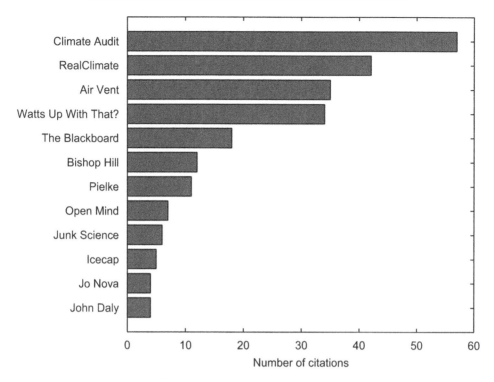

**Figure 1.** Frequency of blog citations in the Reader Background thread at the "Air Vent" blog.

Blogs that can be regarded as in the "lukewarmer" camp include "The Blackboard," with 18 citations, and the blogs run by Roger Pielke Sr. and Jr. (combined into one grouping here, since this was done by some commenters), with 11.

The second most cited blog, with 42 references, is "RealClimate," a long-running blog promoting climate science run by a team of climate scientists. Comments from skeptics are critical of this blog, and many imply that reading it may have been a factor leading to skepticism. Some of these comments say that they were concerned by "RealClimate's" arrogant or dismissive tone, or hostility toward those who disagreed with them. Others report that questions raised were not answered, or in some cases censored. Another blog promoting climate science, "Open Mind," is mentioned seven times, with similar critical comments. Several individuals report that when they started looking into the climate change question, they started reading these blogs but were put off by their style and turned instead to the skeptical blogs.

*Other factors*

Many other factors leading to skepticism are mentioned in the Reader Background thread, occurring less frequently than those listed above. Nine individuals say that they read parts of the reports published by the Intergovernmental Panel on Climate

Change (IPCC) and this led to skepticism (in some cases, this was the apparent endorsement of the "hockey stick" by the IPCC). The issue of some climate scientists being reluctant to release raw data or details of their methodology is raised by six commenters.

Another general area of influence is the media, books, and films. Three skeptics mention the influence of *Sunday Telegraph* reporter Christopher Booker, but newspapers do not appear to be a major factor. There are a few suggestions that exaggerated newspaper reports on global warming may have acted as a spark to ignite skeptical views, and a similarly small number of comments showing that news reports alerted readers to skeptical viewpoints. Books are mentioned eight times but no book appears more than once. Skeptical organizations are mentioned rarely; there is only one reference to the Heartland Institute. The film "An Inconvenient Truth," produced in the USA by Al Gore, is referred to by eight commenters. Two of these appear to have found the film convincing, but the other six indicate that the film instigated or enhanced their skepticism by making exaggerated claims about the science of climate change. The skeptical film "The Great Global Warming Swindle," produced in the UK for Channel 4, is mentioned three times as a factor leading to climate skepticism. A recent paper on the influence of films has found that skeptical films tend to have a greater influence on viewers' opinions than films aimed at increasing climate concern (Greitemeyer, 2013).

*Two similar blog threads*

The "Air Vent" blog's "Reader Background" thread, analyzed in detail above, was the first climate skeptic blog to encourage readers to describe their background and their views on climate science. Since then, two other blogs have carried out similar exercises, briefly discussed in this section.

The "Climate Etc." blog (http://judithcurry.com/) run by Professor Judith Curry, a climate scientist at the Georgia Institute of Technology, launched a "Denizens" thread in November 2010, following the example set by the "Air Vent" blog seven months earlier. There are well over 200 responses on this thread, illustrating similar themes, with a strong science background, concerns with overconfidence and exaggeration, and comments saying that skepticism increased as the issue was studied further, sometimes as a result of politicization of the debate, or due to climategate. The overall view of commenters on this blog is less skeptical than those at the "Air Vent", leaning more toward the "lukewarmer" standpoint. There is, however, a significant overlap between commenters on the two threads, so they cannot be analyzed as two independent data-sets.

More recently, in July 2013 the "Watts Up With That?" blog hosted a guest post entitled "My personal path to catastrophic AGW skepticism" (http://wattsupwiththat. com/2013/07/25/my-personal-path-to-catastrophic-agwskepticism/) where one individual wrote an essay on his background in science and engineering, his initial uncertainty and then increasing doubt, brought on in part by climategate and the influence of skeptic blogs. He encouraged other readers to contribute their experiences, generating over 600 comments. On this thread, one commenter attempted to summarize the main themes, noting the strong science and engineering background; doubts about computer models; the backfiring effect of some prominent climate scientists and blogs; the personal experience of previous apocalyptic warnings; the (relatively rare) tendency for some individuals toward inherent skepticism; and the reluctance of those with right-wing or libertarian views to accept the degree of state control implied by mitigation measures. These points are very similar to those noted in the "Air Vent" thread.

## Discussion and Conclusions

The Internet, in particular via blogs and social media, provides a vast data source for opinion on climate change that has so far been relatively untapped by researchers studying climate communication. Investigation of such resources needs to be carried out carefully, and a "black box" approach, merely analyzing frequency of words automatically, may give misleading results; for example, a particular blog may have a very different effect on different individuals, increasing climate concern for some while decreasing it for others. Inevitably, this makes analysis of such sources a subjective process.

This paper has focused on one particularly informative comment thread on the "Air Vent" blog giving insight into the views of one community of climate skeptics. It is important to acknowledge that this is inevitably a biased sample, influenced by the style of the chosen blog. In this case, the main topics of the blog are technical, scientific aspects of the climate debate, but in this regard the blog is typical of most climate skeptic blogs (Sharman, 2014).

There are a number of possible problems with such an analysis. Foremost among these is the point made above that this is a self-selected sample of individuals, who are certainly not representative of the general public. Nevertheless, this process may shed some useful light on the thinking of skeptical voices in the blogosphere, and in particular, the key factors that may have led them to adopting a skeptical position. Other caveats are that it has been assumed that all comments are genuine and can be taken at face value; there is also the possibility that later comments have been influenced by earlier ones, so the comments may not be completely independent.

Bearing these potential concerns in mind, the following summary conclusions can be drawn regarding the 154 individuals on the Reader Background thread.

- Academic and scientific qualifications are generally high, with 26% reporting a Ph.D. and an additional 46% having a degree of some form. The actual percentages may be higher than this if some individuals did not explicitly mention their qualifications, or lower if some exaggerated theirs. Professional backgrounds are predominantly in technical or engineering fields.
- A broad range of views are expressed, ranging from a "lukewarm" position, that warming will be modest and not a serious problem (around 15–20%) to

a much stronger view that climate change is a scam or a fraud (only around 10%).

- A significant proportion (about 27%) indicates that they were converted to climate skepticism from a previous position of acceptance of climate change.
- Motives for skepticism include the view that claims regarding climate change are often overstated, which in some cases is associated with personal experience of previous exaggerated scares. Blogs that aim to promote climate science can backfire, as they can be seen as overconfident or lacking in objectivity, leading to a potential loss of trust. The main concern of this community of skeptics is with the quality of the science, focusing on issues such as statistics, data handling, and reliance on models, with the hockey stick picture acting as the icon for the dispute. The climategate incident was not a major opinion-forming factor for this group, perhaps because they had formed their opinions before this took place.
- Politics is a significant factor, either through the political views of the individual (which typically lean more toward libertarianism than conservatism) or through the view that those who express concern over climate change may be politically motivated.

Of course, it is very possible to take issue with some of these points. For example, if a scientist appears to make an arrogant statement regarding some aspect of science, it does not follow that the underlying science is wrong (this particular point was in fact discussed on the thread). However, the aim here is merely to observe and present the arguments given by skeptics, not to attempt to address them.

The ironic point that increasingly dire messaging about climate change may encourage skepticism is supported by the work of Feinberg and Willer (2011), and also by Bashir et al. (2013), who found that environmental messages can backfire among those who have a negative view of activists. Similarly, work by Hobson and Niemeyer (2013) found that it is difficult to dispel climate skepticism by subjecting skeptical volunteers to "climate scenarios," and that some became more dogmatic in their skepticism when treated in this way. These results are consistent with the comments studied here.

An important question to be addressed by future research is whether these issues raised by those who comment on climate skeptic blogs translate to the opinions of the general public. Comparing the results obtained here with those obtained from interviews with the general public by Capstick and Pidgeon (2013) shows some similarities, such as doubts about the quality of the science, doubts over the accuracy of predictions, memories of previous scares, and concerns about media exaggeration. Another question is whether the presence of a minority of skeptical opinion in the public debate has a significant impact on public support for the implementation of policy; the work of Aklin and Urpelainen (2013) suggests that it does.

Another emerging issue is the rapidly changing online environment, and in particular the rise in the use of Twitter in the climate debate, which may be to some extent usurping the role of blogs. Two recent papers (Kirilenko & Stepchenkova, 2014;

Pearce, Holmberg, Hellsten, & Nerlich, 2014) have begun to investigate this extensive source of data on the public face of the climate debate.

## Acknowledgement

I am grateful to the anonymous reviewers for their helpful comments on earlier versions of this paper.

## References

Aklin, M., & Urpelainen, J. (2013). Perceptions of scientific dissent undermine public support for environmental policy. *Environmental Science & Policy, 38,* 173–177. doi:10.1016/j.envsci.2013.10.006

Anderegg, W. R. L., & Goldsmith, G. R. (2014). Public interest in climate change over the past decade and the effects of the 'climategate' media event. *Environmental Research Letters, 9,* 054005. doi:10.1088/1748-9326/9/5/054005

Bashir, N. Y., Lockwood, P., Chasteen, A. L., Nadolny, D., & Noyes, I. (2013). The ironic impact of activists: Negative stereotypes reduce social change influence. *European Journal of Social Psychology, 43,* 614–626. doi:10.1002/ejsp.1983

Brulle, R. J., Carmichael, J., & Jenkins, J. C. (2012). Shifting public opinion on climate change: An empirical assessment of factors influencing concern over climate change in the US, 2002–2010. *Climatic Change, 114,* 169–188. doi:10.1007/s10584-012-0403-y

Capstick, S. B., & Pidgeon, N. F. (2014). What is climate change scepticism? Examination of the concept using a mixed methods study of the UK public. *Global Environmental Change, 24,* 389–401. doi:10.1016/j.gloenvcha.2013.08.012

Corner, A., Markowitz, E., & Pidgeon, N. (2014). Public engagement with climate change: The role of human values. *Wiley Interdisciplinary Reviews: Climate Change, 5,* 411–422. doi:10.1002/wcc.269

Engels, A., Hüther, O., Schäfer, M., & Held, H. (2013). Public climate-change skepticism, energy preferences and political participation. *Global Environmental Change, 23,* 1018–1027. doi:10.1016/j.gloenvcha.2013.05.008

Feinberg, M., & Willer, R. (2011). Apocalypse soon? Dire messages reduce belief in global warming by contradicting just-world beliefs. *Psychological Science, 22*(1), 34–38. doi:10.1177/0956797610391911

Greitemeyer, T. (2013). Beware of climate change skeptic films. *Journal of Environmental Psychology, 35,* 105–109. doi:10.1016/j.jenvp.2013.06.002

Grundmann, R. (2013). "Climategate" and the scientific ethos. *Science, Technology & Human Values, 38,* 67–93.

Grundmann, R., & Scott, M. (2012). Disputed climate science in the media: Do countries matter? *Public Understanding of Science, 23,* 220–235. doi:10.1177/0963662512467732

Hagen, B. (2013). *Public perceptions of climate change: Risk, trust, and policy* (Unpublished doctoral dissertation). Arizona State University, Phoenix.

Hmielowski, J. D., Feldman, L., Myers, T. A., Leiserowitz, A., & Maibach, E. (2013). An attack on science? Media use, trust in scientists, and perceptions of global warming. *Public Understanding of Science, 23,* 866–883. doi:10.1177/0963662513480091

Hobson, K., & Niemeyer, S. (2013). "What sceptics believe": The effects of information and deliberation on climate change scepticism. *Public Understanding of Science, 22,* 396–412. doi:10.1177/0963662511430459

Kahan, D. M., Peters, E., Wittlin, M., Slovic, P., Ouellette, L. L., Braman, D., & Mandel, G. (2012). The polarizing impact of science literacy and numeracy on perceived climate change risks. *Nature Climate Change, 2*, 732–735. doi:10.1038/nclimate1547

Kirilenko, A. P., & Stepchenkova, S. O. (2014). Public microblogging on climate change: One year of Twitter worldwide. *Global Environmental Change, 26*, 171–182. doi:10.1016/j.gloenvcha.2014. 02.008

Koteyko, N., Jaspal, R., & Nerlich, B. (2012). Climate change and 'climategate' in online reader comments: A mixed methods study. *The Geographical Journal, 179*(1), 74–86. doi:10.1111/ j.1475-4959.2012.00479.x

Krosnick, J. A., & MacInnis, B. (2010). Frequent viewers of Fox News are less likely to accept scientists' views of global warming. Report for The Woods Institute for the Environment. http://woods.stanford.edu/docs/surveys/Global-Warming-Fox-News.pdf

Lahsen, M. (2013). Anatomy of dissent: A cultural analysis of climate skepticism. *American Behavioral Scientist, 57*, 732–753.

Mann, M. E. (2012). *The hockey stick and the climate wars*. New York, NY: Columbia University Press.

McIntyre, S., & McKitrick, R. (2005). Hockey sticks, principal components, and spurious significance. *Geophysical Research Letters, 32*, L03710.

Montford, A. W. (2010). *The hockey stick illusion: Climategate and the corruption of science*. London: Stacey International.

Montford, A. W. (2012). *Hiding the decline: A history of the climategate affair*. Milnathort: Anglosphere Books.

O'Donnell, R., Lewis, N., McIntyre, S., & Condon, J. (2011). Improved methods for PCA-based reconstructions: Case study using the Steig et al. (2009) Antarctic temperature reconstruction. *Journal of Climate, 24*, 2099–2115.

Painter, J., & Ashe, T. (2012). Cross-national comparison of the presence of climate scepticism in the print media in six countries, 2007–10. *Environmental Research Letters, 7*(4), 044005. doi:10.1088/1748-9326/7/4/044005

Pearce, W., Holmberg, K., Hellsten, I., Nerlich, B. (2014). Climate change on Twitter: Topics, communities and conversations about the 2013 IPCC Working Group 1 report. *PLoS One, 9*, e94785. doi:10.1371/journal.pone.0094785.t009

Pidgeon, N. (2012). Climate change risk perception and communication: Addressing a critical moment? *Risk Analysis, 32*, 951–956. doi:10.1111/j.1539-6924.2012.01856.x

Poortinga, W., Spence, A., Whitmarsh, L., Capstick, S., & Pidgeon, N. F. (2011). Uncertain climate: An investigation into public scepticism about anthropogenic climate change. *Global Environmental Change, 21*, 1015–1024. doi:10.1016/j.gloenvcha.2011.03.001

Santana, A. D. (2013). Virtuous or vitriolic: The effect of anonymity on civility in online newspaper reader comment boards. *Journalism Practice, 8*(1), 18–33. doi:10.1080/17512786.2013.813194

Sharman, A. (2014). Mapping the climate sceptical blogosphere. *Global Environmental Change, 26*, 159–170. doi:10.1016/j.gloenvcha.2014.03.003

Smith, N., & Leiserowitz, A. (2012). The rise of global warming skepticism: Exploring affective image associations in the United States over time. *Risk Analysis, 32*, 1021–1032. doi:10.1111/ j.1539-6924.2012.01801.x

Whitmarsh, L. (2011). Scepticism and uncertainty about climate change: Dimensions, determinants and change over time. *Global Environmental Change, 21*, 690–700. doi:10.1016/j. gloenvcha.2011.01.016

# Structure and Content of the Discourse on Climate Change in the Blogosphere: The Big Picture

Dag Elgesem, Lubos Steskal & Nicholas Diakopoulos

*Based on the texts of 1.3 million blog posts and the structure of the links between the blogs in which these posts appeared, this study presents an analysis of the discourse on climate change in the English-language blogosphere. Our approach combines community detection with probabilistic topic modeling to show how topics related to climate change are discussed across various parts of the blogosphere. We find that there is one community of predominantly climate skeptical blogs but several accepter communities. The topic analysis reveals a series of issues that are characteristic of the climate change discourse in the blogosphere. Two topics, one related to climate change science and one related to climate change politics, are particularly important for characterizing the discourse. We also find that the distribution of topics over the communities cuts across the divide between skeptics and non-skeptics (accepters) and that there are differences in the patterns of interactions between the skeptics and different groups of accepters.*

## Introduction

The discussion of climate change is highly polarized (Dunlap & McCright, 2011; Hoggan, 2009; Mann, 2012; Washington & Cook, 2011), and the struggle has been particularly fierce in the blogosphere (Schäfer, 2012). The blogosphere has been a crucial outlet for climate change skeptics; for example, in the so-called "climategate" affair blogs played a key role in distributing the hacked e-mails (Hoggan, 2009; Pearce, 2010). Another, more recent illustration of the blogosphere's importance to the community of skeptics is their campaign to nominate the skeptical blog "WattsUpWithThat.com" for the yearly Best Blog Award (the Bloggies competition), resulting in the blog's selection as "Best Weblog of The year" in 2013 as well as the "Best Science Blog" for the third time in a row.[1] A number of science blogs chose to withdraw from the competition in protest at the process.[2] But skeptical voices are by no means the only ones in the blogosphere to be heard on the issue of climate change (see, e.g., DeSmogBlog.com, RealClimate.org, SkepticalScience.com).

There is extensive research on climate change coverage in the mainstream media (for overviews, see Boykoff, 2011; Schäfer, 2012), and an important recent study by Sharman (2014) investigates the structure and content of the climate skeptical blogosphere. Nevertheless, we are not aware of any attempts to chart the entire structure of the climate change blogosphere and the topics that get the attention of the bloggers in the different parts of this network. The present research aims to fill this gap by, first, mapping the hyperlink structure of the climate change blogs, then identifying what topics are discussed in the blogs, and, finally, showing which topics characterize the discussion in the various parts of the network. We address three research questions:

(1) How is the hyperlink network of (English) climate change blogs structured?
(2) How does this structure relate to the differences of opinion, as well as the presumed divide, between those bloggers who accept the consensus view on anthropogenic warming and those bloggers who are skeptical of this view?
(3) How is the topic of climate change discussed in the different communities in the blogging network?

As a first step in our investigation, we crawled approximately 3000 English-language blogs discussing climate change, downloaded their links, and extracted the texts of some 1.3 million blog posts. In a second step, we undertook a community detection analysis of the network and visualized it. We then manually classified the blogs as climate skeptics, accepters, or neutral. Afterwards, we subjected the corpus of texts from the blog posts to a probabilistic topic analysis to generate a model of the corpus topics. Upon presenting the results of these analyses, we then show how the topics are distributed over the communities in the climate change blogosphere. Finally, on the basis of the patterns that emerge, we discuss what inferences can be made about the relationships between the structure of links and the topics.

## Overview of the Literature

The media coverage of climate change varies considerably between countries and over time (Boykoff, 2011) with respect to the release of Intergovernmental Panel on Climate Change (IPCC) reports and the yearly Conference of the Parties (COP) meetings (Eide & Kunelius, 2013). Analyses have shown that media with a conservative profile, versus those with liberal or left-leaning positions, give more room to skeptical voices (Painter, 2011). There are also differences between countries in the styles of climate change coverage (Eide & Kunelius, 2013) and in the presence of skeptical voices in the media (Boykoff, 2011; Painter, 2011).

Political affiliation has been shown to be a strong predictor of attitude to climate change issues, with conservatives, versus liberals and leftists, as more likely to be climate skeptics. And it has been documented that conservative political organizations have actively lobbied against legislation to mitigate global warming (Dunlap & McCright, 2011; Oreskes & Conway, 2010).

The debate between the skeptical side and those that represent the majority view is polarized and characterized by hostility (Hoffman, 2011; Washington & Cook, 2011). This is also reflected in the tone of language used in the debates. In relation to the climategate affair, this polarization was driven to new heights with accusations of fraud and scientific misconduct (Pearce, 2010). An interesting recent study of the skeptical discourse online is Koteyko, Jaspal, and Nehrich (2012), which analyses data from readers' comments on online UK tabloids articles about the climategate affair. Using methods from corpus linguistics, they identify characteristic patterns in the negative representation that skeptics give of climate scientists, where the skeptical discourse in these forums is often strongly pejorative.

The term "skeptic" is in itself controversial. Washington and Cook (2011) argue that the label "skeptic" is misleading because it implies that non-skeptical climate scientists are not properly abiding by the scientific norm of rational skepticism. For the authors, those who oppose the theory of anthropogenic climate change should instead be called deniers. Still, in this study we will refer to those who accept the theory of anthropogenic climate change as "climate accepters" and those who reject this theory as "climate skeptics."

Rahmstorf (2004) distinguishes between three types of skeptics: trend skeptics who question global warning, attribution skeptics who question that human activity has significant effects on the climate, and impact skeptics who question the negative consequences from climate change. Sharman (2014) distinguishes between skepticism based on arguments against the science and skeptical arguments directed at political measures aimed at climate change mitigation. In her highly relevant study of skeptical climate change blogs, she finds that the most central skeptical blogs are largely concerned with climate science issues.

With respect to the methods used, several studies have analyzed the manifestation of polarization of political opinions in the online public sphere. Adamic and Glance (2005) charted the linking practices of the top Republican and Democrat blogs and found a strong tendency on both sides to link to other blogs within the same political community but not to the other side. The authors' illustration of a divided political blogosphere in the USA has become an icon of the polarized web. Sunstein (2006, 2007) finds evidence of the same tendency in the linking practices of the websites of political organizations and interprets that tendency as sign of a polarized public sphere. The same phenomenon has been found with communication in social media.

Himelboim, McCreery, and Smith (2013) identify clusters of people who use Twitter to communicate about politically charged topics, including climate change, and show that most of the clusters are quite politically homogenous. Adamic (2008) suggests that a possible explanation for why bloggers are reluctant to link to blogs with which they disagree is that by linking they would give attention and prominence to opposing views. Rogers and Marres (2000) chart the linking practices of websites of industry, government, and activist organizations concerned with climate change and find systematic differences in their linking practices. For example, they found few reciprocal links between industry and activist organizations and suggest that the absence of links in this case is a sign of a lack of recognition of the other party in the climate change discussion.

Several interesting approaches to statistical models of linking practices have been recently suggested. Gonzales-Bailon (2009), Shumate and Dewitt (2008), Ackland and O'Neill (2011), and Lusher and Ackland (2011) analyze the linking patterns among social movement websites and distinguish between links formed on the basis of content similarity (homophily) and links that must be explained with reference to the network's structural patterns. Ackland and O'Neill (2011) also provide a description of their network of environmental websites at the textual level and thereby are able to provide a richer context for interpreting the linking practices. They find that the texts on different webpages often contain the same key terms, even if the websites share few hyperlinks. Still, the exact relationship between the structure of hyperlinks and the relationships at the textual level remain unclear.

## Methods

In this section we explain our methods, with an especially detailed focus on explaining probabilistic topic detection (LDA analysis).

### Sampling, crawling, and extraction

One methodological goal was to maximize our collection of English-language, climate change blogs. A blog is a webpage with articles—blog posts—organized in a chronologically reversed order of their publication. A blog is connected to other webpages via hyperlinks from the blog's stable frame—blog rolls—or from links embedded in the blog post's text. We started with five seed blogs that were chosen because they were well connected in the climate change blogosphere and represented differing viewpoints in the climate debate. We harvested all posts from the seed blogs and extracted key terms from them, which were then used to determine topical relevance in our crawl. First, a frequency sorted word list was scanned to identify frequent words that were typical of the domain, resulting in terms like *climate, global, carbon, emissions, temperature, sea, solar, greenhouse,* etc. Afterwards, lists of the *n*-grams (word sequence length of 2 <= *n* <=5) containing each of these words were extracted and scanned for frequently occurring terms, e.g. *climate change, climate science, carbon dioxide, emissions trading, sea level.* We executed a breadth-first crawl from the seed blogs, with the criteria that for a blog to be harvested it must be in English and have at least one key term appearing somewhere on its homepage. The crawl followed links from the homepage of each blog. If the next blog visited did not match our topic criteria, the crawl went one step further from the blog roll. Because of the effort needed to adapt programs to parse content from different blog platforms, we limited the crawl to Wordpress and Blogspot blogs. By searching blogs for "climate change" and "global warming" with different search engines, it appeared that there were roughly five times as many hits on these two platforms as on all others combined, which suggested that we would capture the bulk of the climate change blogs by crawling these platforms.

In a crawl carried out between June and September 2012, we downloaded the complete content of about 3000 English-language blogs (about 1.3 million posts) that included posts mentioning "climate change" or a related term. From the downloaded html files we extracted the text content of each post by using the Alchemy API[3] and stored this in a MongoDB database, extracted links between blogs, and stored these in a Neo4j graph database.

## Community detection

Although the task of community detection is a well-studied challenge, the notion of what a community is has no single definition and thus no single method for addressing this challenge. Moreover, there is a plethora of methods for finding an underlying structure in a network (Fortunato & Barthelemy, 2007). This implies that even though these methods address the same problem when defined loosely, their outcomes might be similar for some input networks but very different for others.

We decided to use the modularity maximization algorithm (Blondel, Guillaume, Lambiotte, & Lefebvre, 2008). Our reasons were threefold: modularity-based algorithms yield good results and belong to the most popular approaches in the literature (Fortunato & Barthelemy, 2007), they are computationally very efficient, and they are already implemented in the tool we used for network analysis (Gephi).

The intuition behind this method is that a community has more internal links than what would be expected in a random graph with the same properties (Newman, 2006). The modularity score of a set of nodes measures to what extent the number of links within this set is larger (or smaller) than expected in a random network with the same degree distribution. The larger the score, the more likely it is that the links were not generated by chance but by a process common for all nodes in the group. A modularity maximization algorithm tries to ascertain a network partition for which the sum of modularity scores for each partition is maximum.

It has been noted, however, that this approach suffers from resolution problems (Fortunato & Barthelemy, 2007) when the size of some of the actual communities is disproportionate to others (Lancichinetti & Fortunato, 2011). Since we are particularly interested in identifying the most dominant communities and we had no prior information on the scale of existing communities, we chose to disregard this problem. That is why we chose a resolution parameter of 1, which is equivalent to the basic modularity optimization problem (Lambiotte, Delvenne, & Barahona, 2009).

## Manual classification

We undertook a manual classification of the 1497 blogs in the seven largest groups in the central part of the graph identified from community detection. The blogs were classified as "accepting" the majority view on anthropogenic global warming (AGW), "skeptical" of this view, or "neutral" to it. The procedure used was as follows: (1) open the blog in the browser, (2) find the posts on the blog tagged with "climate change" either via the blog's search function or by clicking on the link to the category, (3) search through the result page for the word "climate," (4) inspect the sentences containing this word and its context, and (5) see if it is possible to identify statements that explicitly express the views of an "accepter" or a "skeptic." If no explicit endorsement of either position could be found, the blog was classified as neutral.

During the classification, we looked at more than one statement if we thought the first statement was not clear enough. The default classification of a blog was "neutral." We then checked if it could be classified as "skeptic": we classified a blog as "skeptic" if it explicitly rejected that global warming is happening (trend skeptic), questioned that human activity has an effect on climate (attribution skeptic), or that climate change has serious consequences (impact skeptic). If the blog was not found to be "skeptic" and expressed concern over, respectively, either global warming, that human activity has an effect on climate change, or the impact of climate change, it was classified as "accepter." Blogs that could not be classified as either "skeptic" or "accepter" were labeled "neutral." In a few cases the blog's policy was to represent both positions; these blogs were also classified as "neutral."

Most of the blogs classified as "skeptic" explicitly rejected mainstream climate science. Blogs using terms like "warmist," "alarmist," or "AGW religion" to characterize mainstream climate scientists or people arguing in favor of measures to mitigate climate change were classified as "skeptic." On the other side, several blogs classified as "accepter" expressed views on what should be done in the fight against climate change, even if it never actually mentioned its causes. For example, blogs that said things like "climate change is the most serious threat facing humanity today" were classified as "accepter." This method allowed us to classify the entire blog on the basis of statements in one or a few blog posts, with the assumption that the blogger is consistent and does not change his or her mind. One researcher performed the initial coding, and inter-coder agreement was then tested by letting a second coder classify a sample of 60 blogs. The agreement was 84.8% and the weighted Cohen's kappa was 0.72, which is considered to be sufficient (Fleiss, Levin, & Paik, 2003).

## Topic detection

To identify the topics spanning over our corpus, we decided to use an unsupervised inference method called Latent Dirichlet Allocation (abbreviated LDA; see Blei, Ng, &

Jordan, 2003; Blei and Lafferty, 2009). LDA is a member of a class of methods for characterizing the hidden structure of a large corpus. It is an unsupervised method, meaning that the analysis is carried out automatically without any human support in the analysis. Furthermore, LDA is not a *classification* but a model of the corpus's topical structure. When the algorithm is run on a set of texts, the first output of the analysis is the number of topics where each topic is represented by a set of words that frequently occur together in the corpus. The basis for interpreting a group of words as a topic is that topics are regularly discussed with the use of a characteristic vocabulary. In some cases, the given group of words consists mainly of function words and does not give meaning as a topic.

In this analysis, the topics themselves are represented as hidden variables and are visible only in the form of groups of words from the corpus. Hence, the only manifested elements are the words from the texts. Note that two important assumptions are built into LDA. One is that a given corpus of texts can be modeled by a fixed number of topics. Hence, the number of topics must be given externally as a parameter to the process, and LDA does not provide an estimate of the optimal number of topics for modeling a given corpus. This is a limitation of the method and calls for experimenting with different numbers of topics. The other assumption that LDA makes is that each topic is present in every document to some degree. A second output from the analysis is therefore a distribution of all topics for each document. In practice a small number of topics, one to three, will make up more than 90% of the topics in a given document. For example, in one of our analyses of the climate change corpus, we ran the LDA analysis with 60 topics as a parameter and thus obtained 60 groups with 19 words in each as a suggested corpus model. Five of these topics contained the words "climate change." The result is displayed in Table 1.

We see that the groups of words can be interpreted as signifying various aspects of the climate change issue, i.e. as various topics related to climate change. We have suggested labels for the topics in the column to the right (Table 1). However, the topics generated by this model are not clearly distinct, and the model makes topical

Table 1. The 5 topics (out of 60) identified by the LDA algorithm containing the words "climate change."

| | Topics | Label |
|---|---|---|
| 1 | temperature data climate surface model global models temperatures change al warming heat effect average period trend time years analysis | Climate change science |
| 2 | climate development change countries water world environmental research international sustainable food resources areas economic land work global issues environment | Climate change politics |
| 3 | climate carbon emissions change global countries china world greenhouse energy gas nations dioxide reduce copenhagen trade international kyoto year | Climate change, international negotiations |
| 4 | climate warming global change science scientists scientific ipcc evidence research years weather report gore world earth dr university scientist | Climate change science, IPCC |
| 5 | ice sea ocean climate years carbon arctic global warming earth water change scientists atmosphere level dioxide study levels rise | Climate change and global warming |

distinctions that are not intuitively clear. In the analysis below we use 20 topics to obtain a less fine-grained model which we argue is more adequate.

To carry out the analysis, we used MALLET, a freely available Java implementation of LDA.[4]

## Results

Our first research question was: How is the hyperlink network of the (English) climate change blogs structured?

We first ran the community detection algorithm on the network we harvested. This yielded 19 various-sized groups, as shown in Figure 1.

The size of areas in the graph is proportional to the number of blogs in the groups. Closeness represents density of links. The manual analysis (see below) revealed that the group of blogs colored with red is the predominantly skeptical community, which is also the largest community, while the rest are dominated by accepters. The group colored in yellow, which is one of the groups dominated by "accepters," is the group with which the skeptical community shares the most links.

Note that the size of the different areas represents the number of blogs in the group, not the number of blog posts. The distance between the nodes represents link density. So, for example, there are more links between the large red group and the

**Figure 1.** The network of climate change blogs, colored by community.

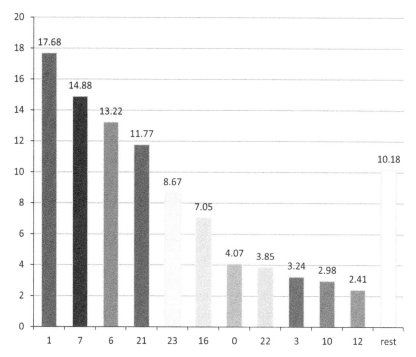

**Figure 2.** The percentage of blogs in each group. The color scheme is the same as in the one used in the graph in Figure 1.

group colored in yellow than between the red group and the purple group to the north. The proportion of blogs in each group is charted in Figure 2.

There are 9 groups with more than 50 blogs, which we have coded as accepting the majority view on climate change, as skeptical of the majority view, or as not expressing a view on the question (neutral). Two of these groups (groups 16 and 21 in Figure 2) consist mainly of blogs that are not classifiable as either skeptics or accepters. These groups are therefore of less interest to our analysis of blogging on climate change. This leaves us with seven groups: 1, 7, 6, 23, 0, 22, and 3.

*Skeptics and accepters*

Our second research question was: How does this structure relate to differences of opinion, as well as the presumed divide, between the bloggers who accept the consensus view on anthropogenic warming and those bloggers who are skeptical of this view?

The results from manually coding the blogs as skeptical, accepters, and neutral yielded the distribution depicted in Figure 3.

As Figure 3 shows, group 1 (colored with red in the graph in Figure 1) is the only one that consists of predominantly skeptical blogs, while the other groups are dominated by blogs that subscribe to the majority view on climate change. Note, however, in all groups there are some dissenters.

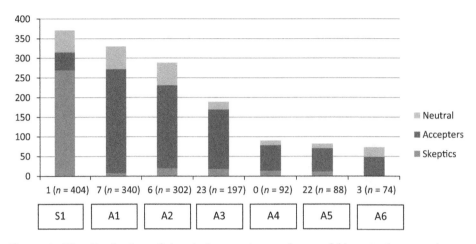

**Figure 3.** The distribution of skeptical, accepting, and neutral blogs in the seven largest among the central groups of blogs concerned with climate change.

In the rest of the study we will refer to the predominantly skeptical group 1 as S1, and in other communities as, respectively, A1 (group 7), A2 (group 6), A3 (group 23), A4 (group 0), A2 (group 22), and A6 (group 3).

We used PageRank (Brin & Page, 1998) as a blog centrality measure, since it is a centrality measure designed specifically for webpages, and it models the number of page views in a random walker model of the user. The advantage over pure degree centrality is that it models the effect of being endorsed by popular blogs. The skeptical blog "WattsUpWithThat.com" (in community S1) came out with the highest PageRank of all the blogs. The blog with the highest PageRank among the blogs in community A1 was "tamino.wordpress.com."

*Topic analysis*

Our third research question was: How is the topic of climate change discussed in the different communities in the blogging network?

We approach this question by using the LDA technique to analyze the texts of the 1.3 million blog posts in our corpus. As noted above, this is an unsupervised technique that automatically generates a set of topics represented by a set of words. However, the number of topics to be found has to be given as a parameter in the algorithm. For this reason, it is necessary to try out the algorithm with different numbers of topics to generate corpus models that can be compared and evaluated. There is no "correct" number of topics that characterize a given corpus: the question is whether the model is useful for the purpose of a given analysis.

Our study aims to compare the discourses in the various parts of the climate change blogosphere, and we ran the LDA algorithm with the number of topics set to 20 and 60. Table 2 reproduces the output from processing with 20 topics. We have also suggested labels for each topic. The labels do not serve any analytic purpose, but

**Table 2**. The outcome of 20-topics analysis in the corpus.

| | Topic | Suggested label |
|---|---|---|
| 0 | 0.10053 obama party president government people political election vote campaign media bush house bill democrats republican don mr time public | US politics |
| 1 | 0.04269 city people road car community street local development urban good building traffic cities public london transportation cars council work | Urban |
| 2 | 0.01052 information site public blog views responsible sites ehs opinions good contained linked understand facts website people material expressed posted | Blogging |
| 3 | 0.05328 water weather river south area north sea service information coast disaster year storm people island high areas west region | Weather |
| 4 | 0.04927 species birds garden food year time trees plants water bird small found plant tree day back good long ve | Wildlife |
| 5 | 0.03707 space science research earth university nasa time system light life years scientists technology researchers field team theory work found | Science and technology |
| 6 | 0.14584 ago time don people day back good ve year years ll make things days home week didn work hours | (incoherent) |
| 7 | 0.07379 state law court public federal government police case act department states county mr rights justice information bill committee office | Government |
| 8 | 0.06804 climate change world countries environmental international development global environment people africa food india resources national economic nations government sustainable | Climate change politics |
| 9 | 0.00278 de da em para os es um se uma na mais dos por como sobre ncia das rio ao | (Portugese) |
| 10 | 0.00682 att och det som en av om inte den med die jag till deals alstom har de du fr | (Swedish) |
| 11 | 0.05081 climate global warming change science data ice carbon temperature years scientists sea earth year emissions temperatures scientific time ipcc | Climate change science |
| 12 | 0.09517 people world life human god time society make good things fact power man political history don book social point | New age |
| 13 | 0.0356 health study medical people children food cancer research care disease environmental public found women risk patients university drug treatment | Health |
| 14 | 0.01038 love life di jesus sl playboy don time age la start change che model joy abortion meaningful universal loving | (incoherent) |
| 15 | 0.06124 blog reading aud site google http news post email www phone information internet min web video uk free posts communication) | (incoherent/ online communication) |
| 16 | 0.06057 energy oil power nuclear gas fuel coal solar production wind carbon electricity industry plant emissions renewable plants technology year | Energy |
| 17 | 0.06955 government money economic tax economy market year people financial billion percent million pay debt bank years world system jobs | Economic policy |
| 18 | 0.05946 war military iraq american world iran government united israel people states security country president bush al nuclear afghanistan forces | US military |
| 19 | 0.00494 de la en el los del las se por es una para hace con le est les al lo | (Spanish) |

The decimal number in front of each topic indicates the relative proportion of the topic in the corpus.

are included to make it easier to reference the topics and to remember the gist of each one.

Not all of these topics make equally good sense. Topics #14 and #15 are both useless and contain terms that should have been filtered out by the stop word list. The Portuguese topic #9, the Swedish topic #10, and the Spanish topic #19 have been identified because they have a different vocabulary than the rest of the corpus and, therefore, are singled out as distinct topics. The documents were included in the corpus in the first place because some of the climate-related words we used in the crawls appeared on the front page of these blogs. The rest of the topics are sufficiently coherent to provide helpful characterizations of different discourse parts that exist in the corpus, and several of them seem to be somehow related to aspects of climate change.

Topics #8 and #11 are of particular interest, of course, since they both contain the words "climate change." They seem to be concerned with two different types of issues related to climate change: topic #11 is a set of terms related to the science of climate change while the words in topic #8 relate more to the politics of climate change (hence the suggested labels). To justify the labels for these topics, we have manually inspected the top 100 documents determined by the algorithm to have the strongest presence of the two topics. Intuitively, the topics of the two groups of blogs are quite distinct, and "climate change science" and "climate change politics" seem to capture the difference.

It seems impossible to give coherent interpretations of topics #6, #14 and #15. This does not invalidate the model but only goes to show that the method could find 15 rather than 20 coherent topics. The fact that the LDA method came up with some groups of frequently co-occurring words that are not intuitively coherent does not mean that the other topics can be used in the discourse analysis. The result of the LDA method is not the final truth about the corpus in question but a suggestion for a model that can be more or less useful. We will argue that the model described in Table 2 provides a useful perspective on the climate change discourses in the blogosphere.

**Validation**

It might be thought to be a problem for the analysis that the the topics #8 and #11 "overlap", since the words "climate change" appear in both topics. Yet that is not the case. First, LDA does not classify documents into mutually exclusive categories but assumes that all topics are present in every document to some degree. Second, LDA denotes topics by identifying groups of words that frequently occur together. That the words "climate change" show up in two topics—i.e. "climate change politics" (#8) and "climate change science" (#11)—only means that they appear frequently in two different contexts. This brings out an interesting and important property of the discourse, and it is therefore a feature, not a defect, of the method, namely its ability to identify different contexts for the use of this central concept.

The appearance of the terms "climate" and "change" in two topics suggests they are used in semantically different ways in two different contexts. To check this we identified for each topic the 100 documents where the topic in question accounted for the largest portion of the text. According to the LDA model, the respective topics accounted for more than 95% of the content in these documents. Hence, if the topics really referred to semantic differences—i.e. something like "climate change science" and "climate change politics"—the differences should be clearly visible in these documents. We tested this by using two methods from corpus linguistics: keyword

**Table 3**. The 18 most frequent keywords from the analysis of the top 100 documents for the topics "climate change science" and "climate change politics."

| Top keywords in the top "climate science" documents | Top keywords in the top 100 "climate politics" documents |
| --- | --- |
| 1 CLIMATE | 1 CLIMATE |
| 2 TEMPERATURE | 2 CHANGE |
| 3 GLOBAL | 3 DEVELOPMENT |
| 4 WARMING | 4 COUNTRIES |
| 5 OCEAN | 5 MOUNTAIN |
| 6 SURFACE | 6 WILL |
| 7 DATA | 7 KNOWLEDGE |
| 8 SEA | 8 ADAPTATION |
| 9 ICE | 9 INTERNATIONAL |
| 10 CHANGE | 10 WORK |
| 11 AL | 11 RESEARCH |
| 12 ET | 12 WATER |
| 13 HEAT | 13 SUSTAINABLE |
| 14 CHANGES | 14 HIMALAYA |
| 15 YEARS | 15 SHARING |
| 16 VARIABILITY | 16 CONFERENCE |
| 17 CARBON | 17 ISSUES |
| 18 MODEL | 18 MEETING |

Reference corpus: The British National Corpus, approximately 100 million words.

analysis of the two groups of documents, and computing the strongest collocates of the words "climate," "change," and "global" in the two groups. In addition, we read through the titles of the 200 documents. Table 3 shows the keyword analysis of the two groups of documents:

We see that the most keywords are quite distinct, and the analysis seems to support our interpretation and the labels.

Our second test was to check what words collocated most strongly with the terms "climate," "change," and "global" in the top 100 documents for "climate change science" and "climate change politics." The result for "climate" is displayed in Table 4. The same test with "change" and "global" showed the same distinction.

In this case we have computed the point-wise mutual information metric of the words with respect to the British National Corpus. The analysis shows a clear difference in the most distinct contexts in which the terms are used in the respective document groups. Moreover, a manual inspection of the document titles supports the assumption that the terms "climate" and "change" are used in semantically distinct contexts in the two groups of texts, thus strengthening our confidence that the model actually has picked out a significant difference in the way climate change is discussed in the blogosphere.

## A polarized discourse

Studies of other media have found that discussion between skeptics and accepters is characterized by unfriendly language and mutual labeling (Koteyko et al., 2012), from

**Table 4.** The top 18 collocates computed with the point-wise mutual information metric.

| Collocates of "CLIMATE" in top "climate science" documents | Collocates of "CLIMATE" in top "climate politics" documents |
| --- | --- |
| 1 DECADAL | 1 INNOVATES |
| 2 VAPOR | 2 HIMALAYA |
| 3 ANTHROPOGENICALLY | 3 DRR |
| 4 INTERDECADAL | 4 HIMALAYAN |
| 5 SMERDON | 5 CANCUN |
| 6 KNMI | 6 ORG |
| 7 FEEDBACKS | 7 MAINSTREAMING |
| 8 SCHEFFER | 8 REDD |
| 9 ANTHROPOGENIC | 9 BIODIVERSITY |
| 10 ERBE | 10 CLIMATE |
| 11 IPCC | 11 MITIGATION |
| 12 GISS | 12 THEMATIC |
| 13 MODELED | 13 IMPACTS |
| 14 KELVINS | 14 STARTER'S |
| 15 THERMOHALINE | 15 CSD |
| 16 SKILLFULLY | 16 ADAPTATION |
| 17 DENIER | 17 DESERTIFICATION |
| 18 MODELING | 18 DURBAN |

Reference corpus: The British National Corpus, approximately 100 million words.

the skeptical side in particular. Our analysis did not produce a topic with these characteristics. However, as part of our corpus analysis, we computed the collocates of the terms "climate," "change," and "global" in all blog posts from each community. We computed what words have the strongest relation of mutual information with these terms as compared with their occurrences in a corpus representing the general language (The British National Corpus). From this analysis, we obtained the top 15 collocates for "climate" in the three largest communities (see Table 5).

As Table 5 shows, the labels that characterize this debate, like "skeptic," "alarmism," "deniers," "contrarians," and others, rank high on the list of strongly informative collocates of "climate." We see that skeptics discuss the labels given to them by accepters (e.g. "deniers") and vice versa (e.g. "alarmism"). These terms are not very frequent, and they do not appear with high regularity (which is why they do not appear in any of the topics), but the collocation analyses show them to be distinctive of how the word "climate" is used in these documents. The same terms are distinctive of the language of the other accepter communities we have discussed. This type of language, which has been found as a characteristic of the polarized debate over climate change, can thus be seen as salient in the climate change discourse in the blogosphere.

## The distribution of topics throughout the groups of blogs

The question now is how these topics are represented at the blog and community levels. In analyzing this we will focus on the seven largest and most central communities. These are the communities containing the blogs that we have manually

**Table 5**. The top 15 collocates around "climate" in communities 1 (skeptic), 23 (accepter), and 7 (accepter) computed with the point-wise mutual information metric.

| Top collocates of "CLIMATE" in the skeptical community S1 | Top collocates of "CLIMATE" in the accepter community A3 | Top collocates of "CLIMATE" in the accepter community A1 |
|---|---|---|
| 1 CLIMATE | 1 DENIERS | 1 POPPIN |
| 2 SKEPTICS | 2 SKEPTICS | 2 DENIERS |
| 3 ALARMISM | 3 CLIMAT | 3 SKEPTICS |
| 4 DENIERS | 4 DECADAL | 4 OBAMA |
| 5 IPCC | 5 CONTRARIANS | 5 WWW |
| 6 DECADAL | 6 OBAMA | 6 EU'S |
| 7 ALARMISTS | 7 NOAA'S | 7 CLIMATE |
| 8 CLIMAT | 8 AGW | 8 YVO |
| 9 CHANGE | 9 WWW | 9 NOAA'S |
| 10 INTERGOVERNMENTAL | 10 DENIER | 10 WILDFIRES |
| 11 OBAMA | 11 CLIMATE | 11 CHANGE'S |
| 12 ANTHROPOGENIC | 12 VAPOR | 12 IPCC |
| 13 AGW | 13 ANTHROPOGENIC | 13 ALARMISM |
| 14 IPCC'S | 14 ALARMISM | 14 PACHAURI |
| 15 WARMING | 15 CONTRARIAN | 15 DENIER |

Reference corpus: The British National Corpus, approximately 100 million words.

coded as skeptics, accepters, and neutral. The LDA algorithm determines the presence of each topic in each document (blog post). To compare the blogs, we have to average the presence of all the topics in all the blog posts in each blog to get the relative "intensity" of each topic in the blog's total output. On the basis of this calculation of the degree of topic presence, we can then move up one level and calculate the presence of the topics in the communities.

Figure 4 shows how the topics "US politics," "climate change politics," "climate change science," "energy," and "economic policy" are distributed over the communities.

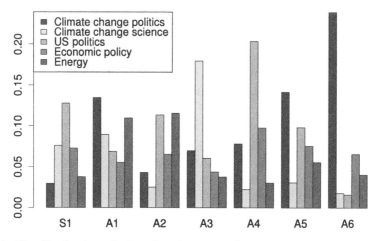

**Figure 4**. The distribution of selected topics across the seven communities.

Note that this is not a representation of the number of blog posts about climate change but the aggregated measure of the presence of the topics in the communities. The graphs show that the topics related to climate change are contextualized in different ways in the different communities. We see that that groups A1, A3, and S1 are the ones with the strongest presence of "climate change science," whereas "climate change politics" is dominating community A6 and strong in A1 and A5.

## Discussion

Blogs are essentially networked texts that are connected both via hyperlinks and via references to other blogs in the texts. To analyze the networks of blogs about climate change we have crawled the hyperlinks of large set of blogs about this topic and harvested the posts published on these blogs. The analysis of the hyperlink structure showed the network to have a number of distinct communities, one of which turned out to be predominantly skeptical while the other communities were dominated by accepters. However, there are also important differences between accepter communities. One distinctive feature is that one of the accepter communities has a much higher level of mutual linking with the skeptical community than the other accepter communities, which suggests a more active engagement with the skeptics.

There are also important differences in the topic profiles between the accepter communities, and the relative emphasis on the scientific and political aspects of climate change varies considerably between the communities. Our analyses show that the central blogs in both the predominantly skeptical community and the central blogs in the accepter community, with which the skeptics engage the most actively, are publishing most about topics related to the science of climate change. This suggests that climate change science is the most central topic in the exchanges between the skeptics and the accepters. The other accepter groups also disagree with the skeptics, but they seem to engage less actively with them, as the links between these communities are sparse. Also, the accepter groups are different in the way they link with each other.

Charting the structure and content of the climate change blogosphere thus reveals distinctive patterns in the structural relationships among these discursive communities, patterns that would be difficult to detect with other methods. We see that the communities are not structured simply on the basis of disagreement between skeptics and accepters but also by various groups of accepters that focus on different aspects of the climate change debate, and these various accepter groups are often not linking to each other. Our analysis of the patterns of topics suggests that the discussion about the science of climate change is distinctively different from the discussion of climate change as a political issue in the blogosphere. We thus find two different types of climate skepticism, and both occur together in the skeptical community. Our analysis also reveals different ways of being an accepter, and suggests many of the accepter blogs are concerned more with issues related to the politics of climate change than with climate science. In addition, the accepter communities differ in the way their writing about climate change brings in other topics like, for example, energy or developmental issues.

Our analysis also finds that not only is the topic "climate change politics" disproportionately prevalent in the skeptical community, but so too is the topic "US politics." The literature finds that climate skepticism often goes together with a conservative political view. While we have not examined the political leanings of the prominent skeptical documents, the present study provides a good starting point for further investigation into this issue. Our findings show that the relationships between the hyperlink structure and the content of climate change discussions are diverse and the disagreement over climate change manifests itself in the structure of links in complex ways. This enables us to go beyond the simple question of whether or not the discussion of climate change is "polarized" and, instead, makes visible a rich set of different relationships between the structure of links and the structure of topics.

## Conclusions

The research reported in this study contributes to the essay of climate change discourse on the Internet in several ways:

- We have, for the first time, charted a major part of the English-language blogs devoted to the issue of climate change and made available for analysis both the hyperlink structure connecting the 3000 blogs and the texts of the 1.3 million blog posts published on them.
- We have identified and analyzed the discursive communities in the climate change blogosphere and found that there is one predominantly skeptical community of bloggers and a number of distinctively different communities of accepters. While there are some studies of the skeptical bloggers little research on the other side exists. The present study has characterized a large portion of the whole of the climate change blogosphere, including the different communities of accepters.
- We have combined network analysis with automated methods for the analysis of the topics of our corpus of blog posts and mapped the distribution of topics over the communities of bloggers. In this way we have characterized the discursive profiles of the different communities and the topical contexts in which climate change is discussed in them. In particular, we have shown the ways in which the emphasis on, respectively, the scientific and the political aspects of climate change differs between the communities.
- We have shown that the patterns of interaction between the communities are diverse and that climate change science is the central topic with the community of accepters that interact the most actively with the skeptical community.
- While our analysis did not produce a distinctively skeptical topic we did find that the skeptical community has a distinct topical profile. We also found that

the name calling often associated with the debate between skeptics and accepters was a salient feature of the discourses across the communities.

## Limitations and future work

The present study has limitations at various levels. First, we have not been able to determine precisely the extent to which the units identified by the community detection algorithm constitute communities in a social sense. It is a plausible assumption that patterns of linking reflect relationships that are in some sense social, but it is difficult to know exactly why people choose to link or not to link with each other. The fact that we find that one community is predominantly skeptic seems correct in the light of what we know about the skeptics as a kind of counter-movement to the science community. Nevertheless, there are accepter blogs and neutral blogs in the skeptical community, but we do not know if this is a sign of diversity in the skeptical camp or whether it is an artifact of the community detection algorithm. A second limitation is that even though we have been able to verify that the two topics related to climate change are distinct in the sense that they point to real semantic differences in the top 100 documents, we have not been able to verify our interpretation of the other documents beyond manual inspection of a selection of documents. A central challenge for future work is to analyze how the discourse on climate change is evolving over time.

## Funding

The research reported in this study has been funded by the Norwegian Research Council's VERDIKT program. Knut Hofland and Andrew Salway of Uni Computing, University of Bergen provided help with the data collection.

## Notes

1. http://2013.bloggi.es/.
2. "Climate skeptics 'capture' the Bloggies' science category." Fred Hickman, *The Guardian*, 1 March 2013.
3. http://www.alchemyapi.com/.
4. http://mallet.cs.umass.edu/.

## References

Ackland, R., & O'Neill, M. (2011). Online collective identity: The case of the environmental movement. *Social Networks, 33*, 177–190. doi:10.1016/j.socnet.2011.03.001

Adamic, L., & Glance, N. (2005). The political blogosphere and the 2004 US election: Divided they blog. KDD'05: Proceedings of the 3rd International Workshop on Link Discovery, Palo Alto, CA, March 4 (pp. 36–43). doi:10.1145/1134271.1134277

Adamic, L. (2008). The social link. In J. Turow & L. Tsui (Eds.), *The hyperlinked society* (pp. 227–248). Ann Arbor: University of Michigan Press.

Blei, D. M., & J. Lafferty, J. (2009). Topic models. In A. N. Srivastava & M. Sahami (Eds.), *Text mining: Classification, clustering, and applications* (pp. 71–93) Boca Raton, FL: Taylor & Francis.

Blei, D. M., Ng, Y., & Jordan, M. I. (2003). Latent Dirichlet allocation. *Journal of Machine Learning research, 3*, 993–1022.

Blondel, V. D., Guillaume, J.-L., Lambiotte, R., & Lefebvre, E. (2008). Fast unfolding of communities in large networks. *Journal of Statistical Mechanics: Theory and Experiment*, 2008, P10008. doi:10.1088/1742-5468/2008/10/P10008

Boykoff, M. T. (2011). *Who speaks for the climate? Making sense of media reporting on climate change.* Cambridge: Cambridge University Press.

Brin, S., & Page, L. (1998). The anatomy of a large-scale hypertextual web search engine. Proceedings of the Seventh International World-Wide Web Conference (WWW 1998), Brisbane, QLD, April 14–18. Retrieved from http://www7.scu.edu.au/

Dunlap, R. E., & McCright, A. M. (2011). Organized climate change denial. In J. Dryzek, R. Norgaard, & D. Schlossberg (Eds.), *The Oxford handbook of climate change and society* (pp. 144–160). Oxford: Oxford University Press.

Eide, E., & Kunelius, R. (Eds). (2013). *Media meets climate.* Gothenburg: Nordicon Press.

Fleiss, J. L., Levin, B., & Paik, M.C. (2003). *Statistical methods for rates and proportions* (3rd ed.). Hoboken, NJ: John Wiley & Sons.

Fortunato, S., & Barthelemy, M. (2007). Resolution limit in community detection. *Proceedings of the National Academy of Sciences, 104*(1), 36–41. doi:10.1073/pnas.0605965104

Gonzales-Bailon, S. (2009). Opening the black box of link formation: Social factors underlying the structure of the web. *Social Networks, 31*, 271–280. doi:10.1016/j.socnet.2009.07.003

Himelboim, I., McCreery, S., & Smith, M. (2013). Birds of a feather tweet together: Integrating network and content analyses to examine cross-ideology exposure on Twitter. *Journal of Computer Mediated Communication, 18*(2), 40–60. doi:10.1111/jcc4.12001

Hoffman, A. (2011). Talking past each other? Cultural framing of skeptical and convinced logics in the climate change debate. *Organization & Environment, 24*(1), 3–33. doi:10.1177/108602 6611404336

Hoggan, J. (2009). *Climate cover up: The crusade to deny global warming.* Vancouver: Greystone Books.

Koteyko, N., Jaspal, R., & Nehrich, B. (2012). Climate change and 'climategate' in online reader comments: A mixed methods study. *The Geographical Journal, 179*(1), 74–86. doi:10.1111/j.1475-4959.2012.00479.x

Lambiotte, R., Delvenne, D.-C., & Barahona, M. (2009). Laplacian dynamics and multiscale modular structure in networks. arXiv preprint arXiv:0812.1770. Retrieved from http://arxiv.org/abs/0812.1770

Lancichinetti, A., & Fortunato, S. (2011). Limits of modularity maximization in community detection. *Physical Review E, 84*(6), 066122. doi:10.1103/PhysRevE.84.066122

Lusher, D., & Ackland, R. (2011). A relational hyperlink analysis of an online social movement. *Journal of Social Structure.* Retrieved from http://www.cmu.edu/joss/content/articles/volume12/Lusher/

Mann, M. (2012). *The Hockey Stick and The Climate Wars.* Columbia University Press.

Newman, M. E. J. (2006). Modularity and community structure in networks. *Proceedings of the National Academy of Sciences, 103*, 8577–8582. doi:10.1073/pnas.0601602103

Oreskes, N., & Conway, E. (2010). *Merchants of doubt: How a handful of scientists obscured the truth on issues from tobacco smoke to global warming.* New York, NY: Bloomsbury.

Painter, J. (2011). *Poles apart. The international reporting on climate scepticism.* Oxford: Reuters Institute for the Study of Journalism.

Pearce, F. (2010). *The climate files: The battle for the truth about global warming.* New York, NY: Random House.

Rahmstorf, S. (2004): *The climate sceptics*. Potsdam: Potsdam Institute for Climate Impact Research.

Rogers, R., & Marres, N. (2000). Landscaping climate change: A mapping technique for understanding science and technology debates on the World Wide Web. *Public Understanding of Science, 9*, 141–163. doi:10.1088/0963-6625/9/2/304

Schäfer, M. (2012). Online communication on climate change and climate politics: A literature review. *WIREs Climate Change, 3*, 527–543.

Sharman, A. (2014). Mapping the climate change blogosphere. *Global Environmental Change, 26*, 159–170. doi:10.1016/j.gloenvcha.2014.03.003

Shumate, M., & Dewitt, L. (2008). The north/south divide in NGO hyperlink networks. *Journal of Computer Mediated Communication, 13*, 405–428.

Sunstein, C. (2006). *Infotopia*. Oxford: Oxford University Press.

Sunstein, C. (2007). *Republic.com 2.0*. Princeton, NJ: Princeton University Press.

Washington, H., & Cook, J. (2011). *Climate change denial*. Oxford: Earthscan.

# Examining User Comments for Deliberative Democracy: A Corpus-driven Analysis of the Climate Change Debate Online

Luke Collins & Brigitte Nerlich

*The public perception of climate change is characterized by heterogeneity, even polarization. Deliberative discussion is regarded by some as key to overcoming polarization and engaging various publics with the complex issue of climate change. In this context, online engagement with news stories is seen as a space for a new "deliberative democratic potential" to emerge. This article examines aspects of deliberation in user comment threads in response to articles on climate change taken from the Guardian. "Deliberation" is understood through the concepts "reciprocity", "topicality", and "argumentation". We demonstrate how corpus analysis can be used to examine the ways in which online debates around climate change may create or deny opportunities for multiple voices and deliberation. Results show that whilst some aspects of online discourse discourage alternative viewpoints and demonstrate "incivility", user comments also show potential for engaging in dialog, and for high levels of interaction.*

## Introduction

For over a decade researchers have supported "deliberation" as part of a decentered democratic process in the implementation of climate change policy (Hayward, 2008; Niemeyer, 2013; Young, 2000). They argue that "[w]here climate change is easily crowded-out in the prevailing nature of political debate, deliberation helps to make salient less tangible and complex dimensions associated with the issue" (Niemeyer, 2013, p. 429). However, there are certain problems with deliberation and democracy in the context of climate change. There is a clash of quite heterogeneous views in online spaces, in particular about the nature of climate change or global warming, its very existence and the validity of scientific statements made about it. There is another clash between what some perceive as scientific uncertainty surrounding climate change (Whitmarsh, 2011), the increasing call for individual members of the public to engage in behavior that mitigates anthropogenic climate change, and a large part of the population perceiving climate change as "low priority" (Upham et al., 2009). Nevertheless, a certain vocal proportion of members of the public engage in online debates about climate change, offering a site to explore how perceptions of climate change as a complex issue are formed, challenged and how they interact with perceptions of science, politics, and economic issues, for example. But do such spaces encourage mediation and deliberative debate? Does engaging in online discussion foster new learning and new understanding in a way that encourages public engagement with the issue of climate change? In the following we shall first review various claims about online debates fostering or inhibiting deliberation and democratic engagement with particular reference to climate change. This will be followed by an examination of one particular discussion thread, demonstrating how corpus analysis can facilitate an examination of features of deliberation in both a quantitative and qualitative way.

## Deliberation

Many researchers have noted that the heterogeneity within climate change discourses is not the product of an information deficit or literacy, but rather based on differences in fundamental beliefs and values (Hulme, 2009; Leiserowitz, Maibach, Roser-Renouf, Smith, & Dawson, 2010; Poortinga, Spence, Whitmarsh, Capstick, & Pidgeon, 2011; Sjöberg, 2003; Slovic & Peters, 1998). As such, different views are not directly linked to scientific evidence and its availability, but rather on individual responses to the same information based on subjective worldviews. This suggests that "deliberation", rather than information or awareness is the key to generating an iterative dialog within the climate change debate. Manosevitch and Walker (2009, p. 8) define "deliberation" as "a political process through which a group of people carefully examines a problem and arrives at a well-reasoned solution after a period of inclusive, respectful consideration of diverse points of view". Wilhelm (1999, p. 156) more succinctly refers to deliberation as "subjecting one's opinion to public scrutiny." Deliberation is the means by which the disparate institutional (in this case, the journalist) and public voices can interact. Positive experiences of deliberation can, it is thought, encourage further engagement. In other words, if the multiplicity of debates around issues such as climate change is shown to create learning outcomes and affect policy for example, continual deliberation is cultivated and the discussion becomes more inclusive.

Ideally, deliberation is based on respecting a diversity of opinions and alternatives in order to arrive at an informed solution and as such, it requires openness: a sense that all contributions can be considered equally. Scheufele, Hardy, Brossard, Waismel-Manor, and Nisbet (2006, p. 730) argue that "[h]eterogeneous discussion leads to a larger 'argument repertoire' and more political knowledge. More political knowledge is positively related to more active participation." This idea has also been understood in relation to "selective exposure," a practice in which users seek opinion-reinforcing content or demonstrate "challenge-aversion" which is seen as problematic and an "anathema to the deliberative perspective" (Freelon, 2013, p. 5; Sunstein, 2009). This idea is supported by Pearce, Holmberg, Hellsten, and Nerlich (2014), who found that Twitter users are more likely to make conversational connections with those who have broadly similar views. There is, however, a possibility that a larger "argument repertoire" increases ambivalence amongst participants. Whitmarsh (2011) argues that the deliberative process is crucial to overcoming the divisive and polarized nature of the climate change debate. Freelon (2013) asserts that a consideration of both deliberation and selective exposure is required in order to account for both the content of online discussion and the ideological relationship between communicators. He states that explorations into the normative aspects of political discourse have typically been understood through principles of "deliberation" but that this has been restrictive. He advocates the application of multi-norm frameworks that go beyond deliberation to include communitarianism and liberal individualism (Freelon, 2010, 2013). Communitarianism refers to the advancement of ideas based on discussion among those who have a shared understanding and who largely do not engage with others except in an adversarial manner. Liberal individualism refers to the practice of self-expression with little mitigation in terms of civility or reciprocity: a monologue within a so-called discussion.

## Online journalism and deliberative democracy

Bowman and Willis (2003) have referred to "citizen journalism" and "participatory journalism" in relation to the growing potential for "user-generated content," which has generated new identities of "prosumers" (producer-consumers) and practices of "produsage" (production-usage; Bruns, 2005; O'Halloran, 2010). Though journalists may remain the "authority" on online content, with online resources we find the greatest potential for that shift from journalism as a "lecture" to a "conversation" (Gillmor, 2003) and the opportunity for discourse as a fundamental principle of democracy (Habermas, 1962/–1989). Reflecting on the impact of the Climategate affair, Holliman (2011, p. 840) observes that:

> journalists are not the only ones who can mine raw online data and generate news. Interested and motivated citizens with sufficient time and access to the web and the requisite skills and competencies in working with scientific data and digital media can assemble as socio-technical networks to generate science news and public debate.

Janssen and Kies (2004) refer to the "cyber-optimists" who assert that the lack of temporal and geographical restrictions, as well as the *online* disinhibition effect (Suler, 2004) encourages greater participation in political issues *online* (Levy, 2002). The very design of *online* spaces facilitates the "multilogue" (Shank, 1993), where unlike spoken discourse a contribution might elicit a number of responses that can be offered at any point in time after the comment. Once a comment has been posted, the conversational floor is open to any of the contributors who can redirect the thread with the content of their *post*. Conversely, the "cyber-pessimists" (Davis, 1999) argue that *online* spaces do not invoke a greater commitment to political debate, rather they undermine the commitment, respect, and sincerity required in deliberative discussion. Furthermore, some scholars have suggested that the freedom and openness associated with *online* discourse has actually led to a fragmentation of public space (Niemeyer, 2012; Sunstein, 2009). Holliman (2011, p. 834) argues that "[w]hilst digital technologies may engender collaboration and collective action, they can also foster disagreement" and found that "many [...] reader comments demonstrated the polarized and sometimes ideologically driven nature of debates about climate change." Painter (2011, p. 5) observes that particularly in the UK and the USA, "climate change has become (to different degrees) more of a politicised issue, which politically polarised print media pick up on and reflect." This apparent polarization suggests an even greater need for more deliberation and raises the question of whether *online* discussion can mediate between the disparate positions adopted and promoted by traditional print media.

Uldam and Askanious (2013, p. 1200) found that comments which followed YouTube posts "did extend the discursive opportunities opened up by the COP15 climate change conference in 2009, facilitating debate between otherwise disparate publics." Hobson and Niemeyer (2012, p. 3) found that "sceptics accounting for themselves in public deliberative settings could indeed potentially foster significant challenges to their beliefs and concerns." And yet, this did not lead to longstanding ideological changes. There are however serious threats to user comment threads generating deliberation, insofar as "the commenting practices on YouTube further impede the emergence of civic cultures because comments frequently are characterized by hostility and do not invite dialogue" (Uldam & Askanius, 2013, p. 1200). Furthermore, "[o]pportunities for user participation in online debate forums are most commonly used to demonstrate opinions in a unidirectional manner rather than to engage in dialogue" (Uldam & Askanius, 2013, p. 1191). This "liberal individualism" is a fundamental aspect of deliberative democracy; however, if users are not engaging with one another then their views become more entrenched: there is little potential for them to develop their perspectives, for mediation or for novel discourses to emerge. Researchers emphasize the need for "more deliberative public engagement techniques in order to break down entrenched camps and seek common societal goals in respect to this complex and morally uncertain issue" (Hulme, 2009; Upham et al., 2009; Whitmarsh, 2011, p. 699).

*User comments*

User comments that appear following news articles published online are one format of discussion that is thought to foster deliberation (Manosevitch & Walker, 2009). User comments are enabled on the websites of all major newspapers in the UK and users need only create a free profile with the website in order to contribute (it is only the *Times* which requires a paid subscription). Discussion threads are "open"—meaning available for comment—for only a short number of days, however they are archived and publically viewable thereafter. Even in the space of a couple of days, articles often attract in excess of 1000 comments and as such, provide a rich resource for the examination of attitudes and opinions around climate change. The amount of data generated poses challenges for researchers to gather a more representative account of such discussions across time, across newspaper websites—even across individual articles. Previous research applying manual content analysis to online user comments has been limited in the scope with which it can examine online debates (Manosevitch & Walker, 2009; Milioni, Vadratsikas, & Papa, 2012). This is particularly true when examining the nuanced ways in which individuals use langue to engage in online debates. In this work, we demonstrate how corpus analysis can aid researchers in pinpointing features of online discussions that can indicate to what extent those discussions are deliberative.

## Methods

Corpus analysis is a systematic and automated process based on the statistical analysis of word frequencies which allows us to process larger data-sets more quickly and more objectively. It is conventionally used to provide a broad overview of the data in reporting keywords and key themes in a data-set. Here, we will identify the features of online discourse that can determine deliberation and how they can be identified. We will also demonstrate how such functions can be developed to identify a sample of key comments from a discussion thread in order for us to conduct a closer analysis of the content of those comments. In order to assess the level of deliberation evident in the data we have identified a number of component aspects of deliberation, based on the literature.

Freelon (2013) identified the following deliberative metrics in his study of online journalism: question asking, opinion justification, and acknowledgment across lines of political difference. Part of a multi-norm framework, he also applied measures of communitarianism (questions, justifications, acknowledgments within lines of polit-ical difference, and calls to political action), and liberal individualism (considering pejorative language and monologic statements). These metrics were applied through content analysis to provide descriptive statistics across a number of online discourse spaces but examples of what constituted each code were not provided. Furthermore, limitations imposed by the codes meant that the researchers were unable to account for alternative normative or deliberative behaviors and some comments could not be properly coded in an "either-or" framework. Wilhelm (1999, p. 156) identifies the following features of the virtual political public sphere:

- Topography: places or spaces in which persons come together to discuss issues, form opinions, and plan action.
- Topicality: the content of discussions or the topics that arise.
- Inclusiveness: notion that everybody has the opportunity to deliberate on policy issues.
- Design: the architecture of the network developed to facilitate/inhibit deliberative discussion.
- Deliberation: subjecting one's opinions to public scrutiny.

Schneider (1997) refers to equality, diversity, reciprocity, and quality, where "quality" is concerned with the topic of discussion. Finally, Hagemann (2002) structures his examination of online Dutch political party lists around questions of: the degree to which the discussion is monopolised by certain members or certain groups of members; reciprocity and the "multilogue"; topicality; and rational argumentation. Based on these studies, our examination of the data was structured around the following topics: reciprocity, topicality, and argumentation, focusing on questions, incivility, and alternative viewpoints.

*Reciprocity*

"Reciprocity" has been defined both in terms of content (Jensen, 2003) but also (somewhat unconvincingly) in terms of structure (Schneider, 1997). Here, reciprocity is examined quantitatively, by looking at the use of specific user names in the discussion. Corpus analysis allows us to examine those usernames referred to most frequently in the discussion, the number of different contributors and the number of contributions made by each user. Research has found that in online spaces purported to facilitate deliberative discussion, there is a tendency for a small number of participants to monopolize the discussion (Jankowski & van Selm, 2000; Schneider, 1997). In the world of social media there are a number of novel ways through which to associate a post with another discussion, group, or individual from the basic hyperlink, to the Twitter "hashtag," or in most online discourse, the use of "@" in front of a moniker. This is one way in which corpus linguistics can provide a quick indication of the level of interactivity between users: by tallying the use of the "@" prefix and with which particular usernames. However, this approach does rely on users using the notation and in the discussion threads examined here it was shown that users would more often simply use the name without the "@" prefix. As such, both the "@" prefix and the use of usernames were considered for evidence of reciprocity. Janssen and Kies (2004) report on research that has defined and used the notion of "reciprocity" as a coding category in content analysis of online discussion forums. Hagemann (2002) examines the content of online posts for levels of (dis-)agreement as an indicator or reciprocity. The systematic examination of reciprocity in the *content* of the comments would require a strict coding strategy, which is not applied here but in the closer examination of a user comment below we show how users might "reciprocate" with one another.

## Topicality

In order to assess the "topicality" of the discussion thread we can utilize semantic annotation to identify key themes in our data. The corpus analysis tool WMatrix (Rayson, 2002) has a built-in semantic categorization function which allocates each word of the data to a category based on its semantic meaning. A full list of the semantic categories can be found here: http://ucrel.lancs.ac.uk/usas/. The software tool is then able to determine which are the most key categories based on a statistical comparison with a normative corpus provided by the British National Corpus as a representation of "normal" language use. This process is systematic and automatic, organizing thousands of words of data into semantic categories in a matter of seconds. By tagging the occurrence of words that make up the key categories in the context of the original discussion thread we can observe the ebb and flow of particular themes throughout the discussion, as well as how those themes converge.

## Incivility

Questions around "civility" are a crucial dimension of the democratic potential of user comments threads in their alienation of users, but may also affect the perception of the actual (scientific) content of the article commented on. The decision made by *Popular Science* to withdraw its comments section (http://www.popsci.com/science/article/2013-09/why-were-shutting-our-comments) was based on a study by Anderson, Brossard, Scheufele, Xenos, and Ladwig (2013) which had shown that the "trolling" and "flaming" associated with the format made readers question the scientific credibility of the original content. Papacharissi (2004) explores the idea of civility and its role in the democratic potential of public discourse, expanding it beyond mere "politeness" theory. "Civility" is grounded in attitudes and beliefs, whereas "politeness" resides in rhetorical style, with name-calling, pejorative speak, and vulgarity deemed "impolite." He argues that "anarchy, individuality, and disagreement, rather than rational accord, lead to true democratic emancipation" (Papacharissi, 2004, p. 266). Therefore, civility goes beyond an interpersonal etiquette and encapsulates a mutual concern for the common good, where disagreement and heterogeneity are fundamental to a public discourse which is critically reflexive. Ultimately, Papacharissi (2004, p. 276) found that "incivility and impoliteness do not dominate online political discussion" and that it was rhetoric rather than incivility that impeded deliberation of the topic at hand in that "[t]he obsession with argumentation skills often led to debates over minute details or even about the principles of argumentation" (Papacharissi, 2004, p. 278). It would seem that users take issue not with what the nature of the argument is, but with the manner in which it is delivered. Anderson et al. (2013) acknowledge that uncivil comments impede the democratic ideal of deliberation and do contribute to the polarization of views around a topic but once again, this is contingent upon individual heuristics, encouraging us to explore "intersubjective positioning" (White, 2003).

As such, identifying "incivility" in a systematic way is problematic, since it can manifest in a variety of language forms and is often indicated in the response, rather than the initial comment. In a closer discourse analysis of the content of discussion we can identify features of vulgarity, pejoration, name-calling, the use of expletives for example, but utilizing corpus analysis to identify "incivility" as a matter of frequency would require establishing a list of specific terms. How corpus analysis was able to aid our examination of incivility in this work was by identifying a sample through which we could conduct a closer discourse analysis.

*Sampling key comments*

The WMatrix corpus analysis tool identifies key categories in the data, which in this work was a user comment discussion thread. By tagging the words of each category in the context of the original thread we can observe the "ebb and flow" of particular themes and where those themes converge. Moreover, we can see which comments incorporated those key themes. Identifying comments that incorporated multiple key themes is one way of extracting a sample for closer analysis. An alternative approach to sampling is offered by Freelon (2013) who analyzed the first 500 characters of comments. However, Freelon (2013, p. 21) did observe a tendency for commenters to punctuate a factual and inquisitive (deliberative) comment with personal attacks and incivility (non-deliberative) in what he termed "deliberative individualism." Thus there is some value in viewing each post as a cohesive unit of analysis and examining the entire comment.

In the discussion thread examined here 17 of 1679 (1.01%) comments included all 10 key categories. Sixty-four comments (3.81%) incorporated nine of the top 10 categories and 159 comments (9.47%) incorporated eight or more categories, suggesting that many of the comments were deliberative in their consideration of the multiple aspects of the climate change debate. As a starting point, this work looked more closely only at those comments that included all 10 categories. Researchers however can be flexible in this criterion depending on the sample size they are looking to extract. A sample of this nature, comments identified through their inclusion key themes, is not going to be representative of the discussion thread as a whole, nor will it incorporate the multiplicity of views in relation to those key themes. But identifying a sample in this way does privilege an assessment of "topicality" in relation to the discussion thread as a whole, since it will contain those semantic categories that have been statistically validated as key to the data. Furthermore, a closer examination of the discourse features of this sample of "key comments" demonstrates that those key categories contain many of the features of language that inform our assessment of the level of deliberation evident in the thread, as is shown below.

## Data

A search was conducted through the *Guardian* website for the term "climate change" from the beginning of its archive up until 31 May 2013. According to the NRS Digital Print and Digital Data survey, the *Guardian* had the largest readership of what were termed the "Quality newspapers" (which included the *Daily Telegraph*, the *Times*, the *Independent and* the *Financial Times*) with 6.4 million visitors each month (http://www.guardian.co.uk/news/datablog/2012/sep/12/digital-newspaper-readerships-national-survey?INTCMP=SRCH). the *Guardian* has enabled readers to make comments online in their "Comment is Free" section since March 2006 and only a week later comments were enabled on all articles across the website (Hermida & Thurman, 2008). From the online archives 30,752 articles were identified through the search term "climate change" however articles making only a passing reference to, for example, "Chris Huhne, Secretary of State for Energy and Climate Change" were excluded and the remainder were ranked by the highest number of user comments. Thirty-three articles from the *Guardian* website elicited 500+ comments, with the highest being 1679 comments. This demonstrates the depth of information available for conducting a longitudinal, cross-case comparison between articles and between newspapers. However, in order to fully demonstrate the analytical methodology, we report only on the article taken from the *Guardian* website with the highest number of comments (1679) written by George Monbiot on the 20 December 2010 entitled "That snow outside is what global warming looks like" (http://www.guardian.co.uk/commentisfree/2010/dec/20/uk-snow-global-warming?INTCMP=SRCH). Discussion threads with the second- and third-highest number of comments (1422 comments and 1295 comments, respectively) [http://www.theguardian.com/commentisfree/cif-green/2009/dec/07/climate-change-denial-industry#start-of-comments; http://www.theguardian.com/commentisfree/cif-green/2009/nov/23/global-warming-leaked-email-climate-scientists] will be referred to for descriptive statistics.

*Moderation*

Fifty-two comments (3.10%) were removed from the discussion thread by a moderator. On the *Guardian* website such comments are replaced with a standard message that also incorporates a link to the site's community standards (http://www.guardian.co.uk/community-standards) and FAQs (http://www.guardian.co.uk/community-faqs). Moderation remains an important consideration for representing the "true" discussion and for liberal individualism, it also has noticeable implications for what users include in their comments. The principles of moderation ensure that the discussion is conducted at a level that reflects the quality and integrity of the newspaper organization as well as protecting contributors from "cyberbullying," but moderation standards are not universal. As such, users are discerning not only about what they write but also where they post it. In this thread one user had six of their seven comments removed, which marginalized their contribution but this may in fact be self-marginalization if the user refuses to alter their discourse to match the standards of the discussion. The thread also features the "CommunityMod": a moderator who actively posts in the discussion and lets users know that they can email them privately to deliberate on the site's moderation practices.

## Analysis

### Reciprocity

Table 1 shows the most prolific contributors to the discussion thread based on the number of comments made, as well as the aggregate percentage of comments made by those users of the discussion thread as a whole and the average word length of their posts. Two users alone were responsible for nearly 10% of the number of comments in this thread; six users accounted for over a fifth of the total comments made. We found similar numbers for the next two discussion threads. There are users who seemingly contribute prolifically to articles around climate change: here, "ElliotCB" appeared in the top 10 contributors for all three discussion threads, contributing over 200 comments in all. The user "gulliver055" appeared in two of three, with a total of 55 comments and many of the others—though not on these lists—were found in the other discussion threads (" Bassireland," "Bioluminescence," "BlueCloud," "heatwave2022," "HypatiaLee," "JBowers,","ShireReeve2," "TruthIsForever," "WheatFromChaff" among others). In the first discussion thread the average post length was 97 words. Though there was great variability between post length, Table 1 shows that the average word length of the most prolific posters was not necessarily above average. This would suggest that the prominence of such users is based on continual engagement with the thread, rather than taking longer "turns." The recurrence of particular users on these threads suggests a certain routine or loyalty in that those who comment on discussion threads in response to articles on climate change on one website are likely to do so again. Given that there are multiple online forums dedicated to the topic of climate change it is not unusual for users to routinely engage with the same site(s). From the content of the posts it is clear that particular users are recognized and their history of comments brought to bear on

**Table 1.** Users who posted the most comments.

| User | Comments | Aggregate % | Comments (after moderation) | Average word length |
| --- | --- | --- | --- | --- |
| JBowers | 99 | 5.90 | 97 | 84 |
| ElliottCB | 62 | 9.59 | 62 | 154 |
| Bluecloud | 61 | 13.22 | 57 | 88 |
| HypatiaLee | 57 | 16.62 | 52 | 106 |
| GeorgeColdwell | 39 | 18.94 | 39 | 125 |
| TruthIsForever | 34 | 20.96 | 25 | 47 |
| andyjr75 | 32 | 22.87 | 29 | 104 |
| Porgythecat | 29 | 24.60 | 29 | 97 |
| Gourdonboy | 28 | 26.27 | 28 | 66 |
| TurningTide | 27 | 27.87 | 27 | 78 |

the current discussion. This "loyalty" not only applies to the newspaper but perhaps more specifically to the journalist.

The *Guardian's* principal journalist on issues to do with climate change is George Monbiot, who wrote the three articles that elicited the highest number of comments on the topic. Commenters express familiarity with his personal stance on climate change issues, indicating that there is something of an in-group: a number of users who are familiar with each other's previous contributions and opinions and who have on more than one occasion been involved in a debate around climate change. This relates to the idea of communitarianism but can create a sense of exclusivity for those who are not as acquainted with the opinions of regular commenters and make it difficult for those less versed in the format or the particulars of the discussion to engage. In fact, in the first thread, though there were 558 different contributors, 363 (65%) only commented once. In the second thread, of 525 contributors, 348 (66%) only commented once and in the third thread, of 548 contributors, 382 (70%) commented just once. This would suggest that the majority of contributors to the discussion are unlikely to fully engage in a dialog with the other contributors since they only make one comment. Other researchers have commented upon the "one-timer effect" (Graham, 2002) and this may be a product of a lack of commitment fostered by the nature of online spaces or an effect of the exclusivity of a particular thread and its participants. We must recognize that one does not have to comment on a discussion thread in order to engage with what has been posted, in fact there is a whole culture of "lurkers" who observe but do not actively engage in online discussion groups. Nevertheless, it is important to determine if this apparent exclusivity is the product of liberal individualism and a monologic type of discourse, whether potential contributors are being excluded because of the nature of the discussion in the thread.

Table 2 shows the usernames that were directly referred to the most in the discussion thread, with or without the "@" prefix. It was shown that those users who were referred to the most were also the users who made the most contributions. This is unsurprising since they were visibly active in the discussion by the number of

**Table 2.** References to user by name.

| Username | Comments made | Referred to | Referred to with "@" | Total |
| --- | --- | --- | --- | --- |
| andyjr75 | 32 | 22 | 2 | 24 |
| Bluecloud | 61 | 19 | 3 | 22 |
| gourdonboy | 28 | 19 | 0 | 19 |
| jbowers | 99 | 14 | 1 | 15 |
| georgecoldwell | 39 | 14 | 1 | 15 |
| macsporan | 22 | 13 | 2 | 15 |
| HypatiaLee | 57 | 12 | 3 | 15 |
| derekbloom | 7 | 9 | 5 | 14 |
| Simongah | 8 | 10 | 2 | 12 |
| euangray | 18 | 6 | 2 | 8 |
| blanketdenial | 2 | 6 | 0 | 6 |

comments they made themselves, provided a greater resource to which to refer. Nevertheless, this shows that the discussion thread was dominated by a handful of users in terms of the comments that were made but also those which were referred to and picked up by other users.

Cavanagh and Dennis (2013, p. 11) found that high posters showed a "marked preference for a dialogical mode of address" which would suggest that such users encourage deliberation, or in the very least that they acknowledge the contributions of others. It was true of the majority of comments in the thread that there were many indicators of references to other comments and commenters in the thread as well as external sources. Of 1679 comments, 1180 (70.3%) made explicit reference to either another user or the author, George Monbiot. The contributors were very much engaging with one another but to determine the effect of this engagement on each users discourse would require a focused longitudinal study. We have provided some indication of the degree to which there is reciprocity but the nature of that interaction requires a closer analysis of the content of the comments.

The sampling method, based on comments incorporating the top 10 key categories, identified 17 "key comments." Eleven of the 17 comments began with reference to another speaker or post: either using the "@" notation, the use of a username or the reproduction of (part of) a post which indicated some basic level of reciprocity. Unsurprisingly, the earlier comments cited the original article and were more likely to address their comments to its author, whereas later comments showed greater interaction between posters as more people became involved. In the example given in Figure 1 the user referred to a specific post in the thread, "As that post said" and referred to a specific user, "As deconvoluter said above." We also observed, "as others have said" and a more general reference to what "A lot of scientists think," demonstrating that posters use both anaphoric and exophoric citations, expanding the discussion beyond the thread.

Thewrongstuff    23 December 2010 11:20. am        8
It's not illogical.
As that post said, AGW is about an increase in average global temperatures that is being caused by human factors/activity (increasing the amount of CO2 in the atmosphere) and nothing else. There is a natural warming and cooling cycle (which is why there have been ice ages and not-ice ages) and the Earth is currently on the warming part of the cycle, but human factors/activity is making that warming more rapid than it would usually be.
As deconvoluter said above, the controversy starts when you start trying to predict how this extra warming/additional energy in the atmosphere will affect the climate and the weather. It's controversial because prediction is an inexact science and something like the precautionary principle is always go to be controversial because you're asking to people to take action to avoid something that might not actually happen.
A lot of scientists think the effect of the extra warming will be bad overall; that is, AGW will lead to *adverse* climate change and weather. AGW and "climate change" are not referring to the same thing (I've no idea how that conflation happened); one is caused by the other, and nobody would be bothered about it (climate change) if it could be proven that the risk of it being bad overall was zero. If the atmosphere could be compared to a pan of boiling water, then AGW would be like turning up the heat a bit and climate change would be the more turbulent boiling and extra steam or whatever. Some parts of the Earth will end up being colder; some parts will end up being hotter, and the weather is more likely to be chaotic as a consequence of more energy being in the system/atmosphere. And colder winters in some parts of the world (followed by hottersummers) are an example of extreme weather.
Extreme (cold) weather won't falsify AGW; as others have said, falsifying AGW would involve proving that average global temperatures are not rising and that somehow proving most of the planet and burning massive quantities of fossiled carbon and hydrocarbons cannot alter the composition of the atmosphere so that the rate of heat transfer from the Earth to space is reduced (the heat isn't trapped; it just takes longer to get out = warming).

**Figure 1.** An example of a key comment. Retrieved from http://www.theguardian.com/ commentisfree/2010/dec/20/uk-snow-global-warming.

**Table 3.** Words representing the top 10 key categories.

| | Semantic category | Words |
|---|---|---|
| 1 | Weather | Climate [430], weather [176], snow [67], snowfall [12], … |
| 2 | Temperature: hot/on fire | Warming [361], warm [50], heat [45], hot [34], hotter[28], … |
| 3 | Science and technology in general | Science [139], scientists [81], scientific [77], scientist [18], … |
| 4 | Evaluation: true | Evidence [102], fact [59], true [40], in_fact [34], facts [29], … |
| 5 | Other proper names | Gaia [21], Nasa [20], guardian [16], CiF [10], google [9], … |
| 6 | Temperature: cold | Cold [140], cooling [50], freezing [23], cooler [10], freezes [10], … |
| 7 | Existing | Is [1066], are [385], be [322], 's [252], was [174], been [87], … |
| 8 | Cause & Effect/Connection | Why [182], effect [43], cause [42], due_to [39], because_of [34], … |
| 9 | Temperature | Temperature [99], temperatures [67], thermometers [5], … |
| 10 | Negative | Not [503], n't [444], no [166], nothing [25], nor [10], none [9], … |

*Topicality*

The key categories for the first discussion thread as identified by the corpus analysis tool are shown in Table 3. Unsurprisingly, the most prolific category was "Weather" which incorporated all uses of the term "climate," as well as words referring to various aspects of weather. There were three separate categories concerned with temperature, which is testament to the notion that discussions about climate change are generally framed as a rise or fall in temperature, incorporating the debate about the misnomer "global warming," The category of "Science and technology" was significant, incorporating all forms of the word "science" and echoing the findings of Koteyko, Jaspal, and Nerlich (2013) in their analysis of user comments taken from articles on climate change published in the *Daily Mail.* Of 1679 comments 1467 (87.4%) made at least one reference to "Weather," "Temperature," or "Science". We also found a preoccupation with "evidence," "facts," and "truth" in the fourth category. The discussion was also characterized by considerations of causality, as users considered the relationship between for example, climate, weather, and temperature through terms such as "due_to," "because_of," and "cause" in the eighth category. A category of terms of negation was the tenth most significant. When considered in relation to the seventh category (which referred to terms of "being," what "is," "was," and has "been") this reflected a tendency in the discussion to refer to what "is" and "what is not." This type of discussion may indicate some level of reciprocity as users respond to claims with counter-claims, but would also suggest that there is little deliberation here since the assertions are delivered in such a matter-of-fact way. The category of "Other proper names" incorporated the acronyms "anthropogenic global warming" (AGW) and "Intergovernmental Panel on Climate Change" which we would expect in a debate around climate change, as well as media companies such as the *Guardian* itself and the BBC. However, the majority of terms in this category were the "handles" of users in the discussion. The "signatures" were removed so the occurrence of a username demonstrated a direct reference by one user to another user's comment, or to the user themselves. The reference to usernames, to some degree an indicator of reciprocity, was common enough in the three discussion threads that it was one of the significant semantic categories in each instance.

### Incivility

As was reported above, 52 (3.10%) of the comments in the first discussion thread were removed by the moderator. Based on the newspaper's guidelines we can only presume that these comments were characterized by "incivility." Examples of vulgarity, pejoration, name-calling, and stereotyping were evident in the sample of key comments but were secondary to a demand for well-reasoned argumentation, as shown in this example:

> If you haven't got a rational scientific explanation for the changes we are experiencing that provides a better fit theory than man made *climate* change, and which you can back up with scientific evidence, then please, SHUT THE FUCK UP.

Freelon (2013) also observed a tendency to punctuate a factual and inquisitive (deliberative) comment with personal attacks and incivility (non-deliberative) in what he termed "deliberative individualism." As a matter of style, we might consider how users capitalize on the impact that a pejorative or expletive statement has (emphasized by the use of capital letters) and seem content enough to punctuate their more reasoned assertions in this way.

### Questions

Examples of questions in the sample of key comments could be understood in relation to justification (as a form of rhetoric) and to deliberation as a matter of inquiry. There were many examples of the use of rhetorical questioning, from the basic "Really?" to indicate doubt; in ridicule, "I mean climate scientists knowing about climate? Who'd have thought it?" or as the pre-cursor to the poster's assertion or justification: "And what do the satellites show? Well." However it was often difficult to determine if the questions posed in this sample were used for rhetorical effect or for genuine inquiry. In one key comment a commenter produced a sequence of seven questions with no clear sense of whether an informative response was required. For example, they asked "if it's not about sea ice then our freezing temperature is not to do with global warming. So, why do warmists tell us it is?," and "Why I am [sic] stupid for following the logic of what AGW supporters are saying?" Certainly in the latter example, this form of questioning could be employed to imply

process developed from the identification of key semantic categories identified key comments, facilitating a closer textual analysis. This sampling allows us to examine the multiple ways in which language can evidence reciprocity and characterize argumentation. The variability in this rhetorical style justifies a closer examination of the context for discourse features and supports a combined quantitative and qualitative approach. Corpus analysis has been shown not only to facilitate that combined approach in its fundamental features of frequency analysis but also in allowing us to extract a smaller sample of comments.

This approach demonstrated that incivility was peripheral to the discussion and that key comments were characterized by more sophisticated argument structures. In response to his observations of a "deliberative individualism"—where deliberative comments are juxtaposed with insulting language and incivility—Freelon (2013, p. 22) suggests that the simple removal of offensive comments would allow the deliberative aspects to "shine through unadulterated." The sampling process shown here did not remove all aspects of incivility but it does privilege more developed comments that would also consider the key themes of discussion. The interaction between key themes showed that they can be thought of as cohesive and interrelated, rather than just appearing in close proximity. Many of the linguistic components that conveyed aspects of deliberation were those very words that formed the key categories: the reference to other usernames demonstrated a level of reciprocity and interaction; the categories of "Existing" and "Negative" located many of the "matter-of-fact" statements that also conveyed reciprocity and liberal individualism; and the "Cause and Effect" category pertained to a level of justification and argumentation.

Corpus analysis provides some of the tools through which the broader interactions of online journalism can be examined (such as username frequency) as well as facilitating a sampling process through which a closer examination of the discourse can take place on a broader scale. To examine in more detail how the deliberative potential of such a format is realized would warrant a focused, longitudinal study on particular contributors to the thread as well as the study of other comment threads following articles on climate change in other newspapers and blogs. We must consider how the readers of the *Guardian* for example might interact differently when compared to other online users. A more sequential analysis of the comments would show how users influence each other's thinking and how certain themes become more prominent in the debate, considering the interactional processes of deliberation, and the ways in which deliberation brings about a change in perspective. Researchers have looked at the effects of deliberation through Deliberative Polls for example, where there are mitigating factors such as the salience of the topic and the potential for individual learning (List, Luskin, Fishkin, & McLean, 2006). Examining the content of the article and the discussion thread would offer some insight into the relationship between the journalist/media and their readership. Using a novel corpus analysis technique combined with closer text analysis this work has shown that the most prolific contributors engaged with other users in the climate change debate and foregrounded well-reasoned argumentation over incivility, offering some evidence of deliberation in online discussion threads related to climate change.

## Acknowledgments

We would like to thank the reviewers and the editorial team for their comments on an earlier draft of this paper.

## Funding

This work was supported by the ESRC-funded project, "Climate Change as a Complex Social Issue: From Greenhouse Effect to Climategate: A systematic study of climate change as a complex social issue," project reference RES-360-25-0068.

## References

Anderson, A. A., Brossard, D., Scheufele, D. A., Xenos, M. A., & Ladwig, P. (2013). The "nasty effect:" Online incivility and risk perceptions of emerging technologies. *Journal of Computer-Mediated Communication, 19*, 373–387. doi:10.1111/jcc4.12009

Bowman, S., & Willis, C. (2003). Introduction to participatory journalism. In S. Bowman & C. Willis (Eds.), *We media: How audiences are shaping the future of news and information.* (Vol. 11(1), pp. 7–8). Reston, VA: The Media Center at the American Press Institute. Retrieved October 23, 2013, from http://www.hypergene.net/wemedia/download/we_media.pdf

Bruns, A. (2005). *Gatewatching: Collaborative online news production.* New York, NY: Peter Lang.

Cavanagh, A., & Dennis, A. (2013). Having your say: The social organisation of online news commentary. *Sociological Research Online, 18*(2), 4–13. doi:10.5153/sro.2919

Davis, R. (1999). *The web of politics.* Oxford: Oxford University Press.

Freelon, D. (2010). Analyzing online political discourse discussion using three models of democratic communication. *New Media & Society, 12*, 1172–1190. doi:10.1177/1461444809357927

Freelon, D. (2013). Discourse architecture, ideology and democratic norms in online political discussion. *New Media & Society,* 1–20. Retrieved October 24, 2014, from http://nms.sagepub.com/content/early/2013/12/02/1461444813513259.full.pdf+html

Gillmor, D. (2003). Moving toward participatory journalism. *Nieman Reports, 57*(3), 79–80.

Graham, T. S. (2002). *The public sphere needs you. Deliberating in online forums: New hope for the public sphere?* (MA dissertation). Amsterdam: The Amsterdam School of Communications Research.

Habermas, J. (1962/1989). *The structural transformation of the public sphere: An inquiry into a category of bourgeois society.* (Thomas Burger, Trans.). Cambridge, MA: The MIT Press.

Hagemann, C. (2002). Participation in and contents of two Dutch political party discussion lists on the Internet. *Javnost–– The Public, 9*(2), 61–76.

Hayward, B. (2008). Let's talk about the weather: Decentering democratic debate about climate change. *Hypatia, 23*(3), 79–98. doi:10.1111/j.1527-2001.2008.tb01206.x

Hermida, A., & Thurman, N. (2008). The integration of user-generated content within professional journalistic frameworks at British newspaper websites. *Journalism Practice, 2*, 343–356. doi:10.1080/17512780802054538

Hobson, K., & Niemeyer, S. (2012). "What sceptics believe": The effects of information and deliberation on climate change scepticism. *Public Understanding of Science, 22*, 396–412. doi:10.1177/0963662511430459

Holliman, R. (2011). Advocacy in the tail: Exploring the implications of "climategate" for science journalism and public debate in the digital age. *Journalism, 12,* 832–846. doi:10.1177/1464884911412707

Hulme, M. (2009). *Why we disagree about climate change.* Cambridge: Cambridge University Press.

Jankowski, N. W., & van Selm, M. (2000). The promise and practice of public debate in cyberspace. In K. L. Hacker & J. van Dijk (Eds.), *Digital democracy: Issues of theory and practice* (pp.149–165). London: Sage.

Janssen, D., & Kies, R. (2004, May 22–23). *Online forums and deliberative democracy: Hypotheses, variables and methodologies.* Paper prepared for the conference on "Empirical approaches to deliberative politics", European University Institute, Florence. Retrieved April 29, 2014, http://edemocracycentre.ch/files/onlineforums.pdf

Jensen, J. L. (2003). Public spheres on the internet: Anarchic or government-sponsored-- a comparison. *Scandinavian Political Studies, 26,* 349–374. doi:10.1111/j.1467-9477.2003.00093.x

Koteyko, N., Jaspal, R., & Nerlich, B. (2013). Climate change and "climategate" in online reader comments: A mixed methods study. *The Geographical Journal, 179*(1), 74–86. doi:10.1111/j.1475-4959.2012.00479.x

Leiserowitz, A., Maibach, E. W., Roser-Renouf, C., Smith, N., & Dawson, E. (2010). *Climategate, public opinion, and the loss of trust* (Working Paper SSRN). Retrieved October 23, 2013, from http://ssrn.com/abstract=1633932

Levy, P. (2002). *Cyberdemocracy.* Paris: Odile Jacob.

List, C., Luskin, R. C., Fishkin, J. S., & McLean, I. (2006). *Deliberation, single-peakedness, and the possibility of meaningful democracy: Evidence from deliberative polls* (PSE Working Papers 1–2006). London: Department of Government, London School of Economics and Political Science. Retrieved October 24, 2014, from http://eprints.lse.ac.uk/20069/1/PSPE_WP1_06_(LSERO).pdf

Manosevitch, E., & Walker, D. (2009, April 17–18). *Reader comments to online opinion journalism: A space of public deliberation.* Paper presented at the 10th International Symposium on Online Journalism, Austin, TX.

Milioni, D., Vadratsikas, K., & Papa, V. (2012). "Their two cents worth": A content analysis of online readers' comments in mainstream news outlets. *Observatorio (OBS*), 6*(3), 21–47.

Niemeyer, S. (2012). *Building the foundations of deliberative democracy: The deliberative person and climate change* (Working Paper 2012/1). Canberra: Australian National University. Retrieved October 22, 2013, from http://deliberativedemocracy.anu.edu.au/sites/default/files/documents/working_papers/2012%201%20Simon%20Niemeyer.pdf

Niemeyer, S. (2013). Democracy and climate change: What can deliberative democracy contribute? *Australian Journal of Politics & History, 59,* 429–448. doi:10.1111/ajph.12025

O'Halloran, K. A. (2010). How to use corpus linguistics in the study of media discourse. In A. O'Keeffe & M. J. McCarthy (Eds.), *The Routledge handbook of corpus linguistics* (pp. 563–577). Abingdon: Routledge.

Painter, J. (2011). *Poles apart: The international reporting of climate scepticism.* Oxford: Reuters Institute for the Study of Journalism.

Papacharissi, Z. (2004). Democracy online: Civility, politeness, and the democratic potential of online political discussion groups. *New Media & Society, 6,* 259–283. doi:10.1177/1461444804041444

Pearce, W., Holmberg, K., Hellsten, I., & Nerlich, B. (2014). Climate change on twitter: Topics, communities and conversations about the 2013 IPCC Working Group 1 report. *PLoS ONE, 9*(4): e94785. doi:10.1371/journal.pone.0094785.t009

Poortinga, W., Spence, A., Whitmarsh, L., Capstick, S., & Pidgeon, N. F. (2011). Uncertain climate: An investigation into public scepticism about anthropogenic climate change. *Global Environmental Change, 21,* 1015–1024. doi:10.1016/j.gloenvcha.2011.03.001

Rayson, P. (2002). *Matrix: A statistical method and software tool for linguistic analysis through corpus comparison* (PhD thesis). Lancaster University, Lancaster, UK.

Scheufele, D. A., Hardy, B. W., Brossard, D., Waismel-Manor, I. S., & Nisbet, E. (2006). Democracy based on difference: Examining the links between structural heterogeneity, heterogeneity of discussion networks, and democratic citizenship. *Journal of Communication, 56,* 728–753. doi:10.1111/j.1460-2466.2006.00317.x

Schneider, S. M. (1997). *Extending the public sphere through computer-mediated communication: Political discussion about abortion in a usenet newsgroup.* (Doctoral thesis). Massachusetts Institute of Technology, Cambridge, MA, USA.

Shank, G. (1993). Abductive multiloguing. The semiotic dynamics of navigating the Net. *The Arachnet Electronic Journal on Virtual Culture, 1*(1). Retrieved August 22, 2014, from http://www.infomotions.com/serials.aejvc/aejvc-v1n01-shank-abductive.txt

Sjöberg, L. (2003). Risk perception is not what it seems: The psychometric paradigm revisited. In K. Andersson (Ed.), *VALDOR Conference 2003* (pp. 14–29). Stockholm: V Aldor.

Slovic, P., & Peters, E. (1998). The importance of worldviews in risk perception. *Risk Decision and Policy, 3,* 165–170. doi:10.1080/135753098348275

Suler, J. (2004). The online disinhibition effect. *Cyberpsychology & Behavior, 7,* 321–326. doi:10.1089/1094931041291295

Sunstein, C. (2009). *Republic.com 2.0.* Oxfordshire: Princeton University Press.

Uldam, J., & Askanius, T. (2013). Online civic cultures: Debating climate change activism on YouTube. *International Journal of Communication, 7,* 1185–1204.

Upham, P., Whitmarsh, L., Poortinga, W., Purdam, K., Darnton, A., McLachlan, C., & Devine-Wright, P. (2009). Public attitudes to environmental change: A selective review of theory and practice. *A research synthesis for the living with environmental change programme.* Swindon: Research Councils. Retrieved October 23, 2013, from http://www.lwec.org.uk/

White, P. R. R. (2003). Beyond modality and hedging: A dialogic view of the language of intersubjective stance. *Text, 23,* 259–284.

Whitmarsh, L. (2011). Scepticism and uncertainty about climate change: Dimensions, determinants and change over time. *Global Environmental Change, 21,* 690–700. doi:10.1016/j.gloenvcha.2011.01.016

Wilhelm, A. G. (1999). Virtual sounding boards: How deliberative is online political discussion? In B. N. Hague & B. D. Loader (Eds.), *Digital democracy. Discourse and decision making in the information age* (pp. 154–178). London: Routledge.

Young, I. M. (2000). *Inclusion and democracy.* Oxford: Oxford University Press.

# Meeting the Climate Change Challenge (MC³): The Role of the Internet in Climate Change Research Dissemination and Knowledge Mobilization

Robert Newell & Ann Dale

*This paper explores the role that Internet and online technologies played in research dissemination and knowledge mobilization in a recent climate change research project, MC³. In addition, the team looked at the potential of online expert-practitioner research collaborations for these purposes. Electronic communication was seen as a key element for creating distributed networks essential to the project and for building new practitioner/research knowledge collaboratives. The paper discusses how online communication strategies and technologies were used for wide dissemination of its research outcomes. MC³'s research dissemination and knowledge mobilization strategies are analyzed, using engagement as the primary measure, to gain insights on the effectiveness and challenges of using Internet-based tools for communicating climate change innovations and actions.*

## Introduction

There is a growing body of evidence that climate change will severely affect the capacity of communities and societies at local, regional, national, and global scales in the future (Easterling et al., 2000; Intergovernmental Panel on Climate Change [IPCC], 2007; Stern, 2006). Addressing climate change, therefore, calls for coordinated efforts across sectors, levels of government, and internationally (Dale, 2001; Dale, Robinson, Herbert, & Shaw, 2013) and the inclusion of multiple stakeholders (Few, Brown, & Tompkins, 2007). Modern online technologies have provided innovative new ways of assisting communities in addressing climate change (Girvetz et al., 2009; Wenkel et al., 2013) and opportunities for involving the broader public in climate action (Biggar & Middleton, 2010). Advancements in online communications also contribute to the building of social capital across political and geographic boundaries and the formation of global networks, referred to by Smith and Smythe (1999, p. 88) as a "global civil society" or "myriad groups and networks of action and knowledge that can ... extend across state borders." We now have the tools and online networks; how can they be used more effectively to accelerate knowledge and action on climate change adaptation and mitigation?

## Meeting the Climate Change Challenge (MC$^3$) Project

The MC$^3$ project examined climate change innovations at the community scale in the Canadian province of British Columbia (BC). The two-year project was launched in September 2011 with the publication of its website (mc-3.ca) and completed in June 2013 with publication of an action agenda for decision-makers. The project involved a tri-university partnership, including researchers from Royal Roads University, the University of British Columbia, and Simon Fraser University, and 12 research partners from the public and private sectors. The principal investigator and research associate from Royal Roads University are the authors of this paper, and lead the Canada Research Chair in Sustainable Community Development program (based out of Royal Roads), hereafter referred to as the CRC program (crcresearch.org).

In addition to being the home province of the researching universities, BC was selected as a focus for MC$^3$ because the province has engaged in a suite of policies and initiatives that have encouraged and enabled local climate action and innovation (Burch, Herbert, & Robinson, 2014; Dale et al., 2013). Among these initiatives is the BC Climate Action Charter, a voluntary instrument established in 2008 for local and regional governments to reduce their carbon emissions (Burch et al., 2014). In addition, the province amended the Local Government Act in 2008 (Bill 27), mandating that local governments include policies and actions in their official community plans that will lead to reductions in greenhouse gas emissions (Berkhout & Westerhoff, 2013). Therefore, albeit local culture and politics can change from community-to-community in BC, the provincial legislative and policy framework, such as the Climate Action Charter and Bill 27, have spurred local climate innovation and action in BC (Berkhout & Westerhoff, 2013), thus allowing the province to serve as a "living laboratory" for examining community climate innovation and action.

The specific objectives of MC$^3$ were two-fold. First using a mixed-methods and contextual, comparative case study approach (Stake, 1995; Yin, 2003), the research aimed to identify and investigate innovative municipal approaches to climate action and document best practices. $MC^3$ researchers developed case studies on 11 BC communities that were identified by the research team and research partners as having implemented (or were in the process of implementing) climate change innovations. The case studies ranged in scale from one of Canada's largest urban centers (Vancouver) to mid- and smaller communities, and they were geographically, politically, and culturally diverse, including a First Nations community (T'Sou-ke) and the province's capital city (Victoria). Case study data were collected through an open-ended, semi-structured interview protocol with local government politicians and staff, practitioners, and civil society leaders until conceptual saturation was reached (Newman, 2003).

The second objective and the focus of the current paper started iteratively from the beginning of the project. This involved knowledge mobilization and research outcome dissemination locally and provincially in order to bootstrap innovation diffusion, optimize local government and provincial partnerships, and share lessons from leading communities taking climate action with those less advanced. This outcome was heavily reliant on Internet communications as it involved disseminating information among geographically dispersed communities, as well as building public profile and commitment to the research outcomes. It involved five channels: an online case study library, online real-time e-Dialogues and LiveChats, social media, peer-to-peer learning exchanges, and traditional academic dissemination (conferences, workshops, peer-reviewed journal articles). $MC^3$ also experimented in knowledge mobilization involving face-to-face interaction, specifically a peer-to-peer learning exchange held at Royal Roads University that brought together provincial community leaders, government staff, and climate innovators. However, the face-to-face initiatives constituted some of the more labor-intensive activities of the project, and the majority of $MC^3$ knowledge mobilization and research dissemination efforts were, thus, Internet-based.

This paper describes outcomes around the latter of these two objectives, assessing $MC^3$'s performance in mobilizing knowledge and disseminating research, and focuses specifically on Internet-based research delivery strategies used by $MC^3$. This is not to discount the value of other communication channels and face-to-face knowledge exchanges; however, as $MC^3$ information sharing relied heavily on online tools, this project has provided a valuable opportunity for evaluating the use of online tools for sharing community climate innovations.

The following sections discuss (in order): the strategies and software used for research dissemination and knowledge mobilization; analysis of how effective they were in dissemination and engaging audiences; and major insights regarding challenges and opportunities in using the online communications tools for diffusing climate change research.

## Internet Knowledge Mobilization and Research Dissemination

With the exception of traditional academic publishing, each of $MC^3$'s five channels had a specific Internet-based delivery method. In order to optimize dissemination and increase engagement in the research, these online tools were used synchronously, meaning they linked and/or referred to one another. In some cases, certain online tools were specifically employed to amplify the reach of other Internet-based resources, for example, the use of social media programs such as Twitter and Facebook to draw in public audiences to e-Dialogues and LiveChats. Figure 1 displays the relationship between the various channels and the intended target audiences. Four of the five channels mentioned above are featured, excluding traditional academic dissemination. Each of the communication channels with respect to online tools and target audiences is discussed in detail below.

## Case studies

The primary research activity of MC$^3$ consisted of developing case studies on 11 BC communities, each of which had implemented (or were in the process of implementing) innovative climate action plans and strategies. Case communities ranged in size, economy type, geography, and urbanness. The case selection criteria purposefully captured a diversity of communities to ensure a comprehensive survey

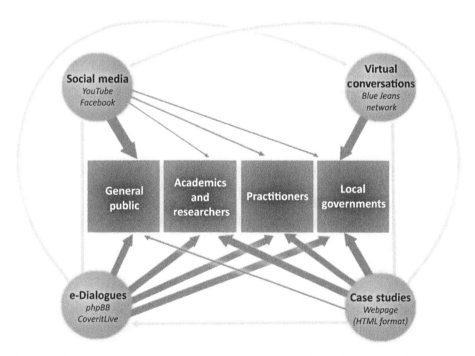

**Figure 1.** Internet-based strategies for mobilizing knowledge and disseminating research. Note: Dark gray arrows identify which audiences are targeted with a specific channel, and weighting of the arrows indicate the level to which a particular audience type was intentional targeted. Light gray areas display relationships between the different online channels, and the specificities of these relationships are identified by arrows.

scan of the challenges and opportunities for implementing climate action in a variety of different types of communities across the province (and country).

MC$^3$ case study research held practical applications, as it served as a method of compiling lessons (capturing both challenges and successes) on how to effectively implement community-scale climate action. Because this research was conducted on a diversity of communities, case studies were used to determine how to implement climate change best practices in other communities of a variety of sizes, urbanness, and so forth. They were created using a format developed through and previously used by the CRC program (http://www.crcresearch.org/community-research-connections/crc-case-studies), designed to present case study material in a clear and intuitive manner, understandable (and thus accessible) by both academics and non-academics. The case study format consists of key aspects of the case provided at the beginning of the study including: sustainable development characteristics, success factors, what worked, what did not work, sources of funding and contact details for the key actor(s) behind the innovation (for those looking for more information). A more academic description of the case study is detailed at the end of the case. The case study tool is online and web-based, encouraging timely research dissemination prior to any academic publishing (i.e., in the same year data was collected) and practical application of the research. This ensures that the research is current, which is essential when examining best practices for climate action innovations and states of implementation, and for speeding the exploitation of knowledge (Kurtz & Snowden, 2003) and knowledge transfer. In addition, building an online case study library presents several opportunities for disseminating information and engaging audiences. For example, since case studies were embedded directly into webpages of the MC$^3$ website (HTML format), this allowed for hyperlinking from the home page, other websites, and social media channels (discussed in further detail below).

Many of case studies featured climate innovations and action conducted through policy, infrastructure, and land-use planning; therefore, in terms of mobilizing knowledge and spreading innovation, local government officials and practitioners working in partnerships with local government comprised an important target audience. In addition, MC$^3$ research identified academic-government partnerships as valuable for advancing climate action (i.e., the partnership between the City of Prince George and University of Northern British Columbia; Newell & King, 2012), and thus, climate policy researchers also formed another important target audience for case studies. Accordingly, dissemination of case studies involved reaching out to researcher and government networks, and used the researcher-government networks formed through the project and previously established through the CRC program. Case study links were shared among participants of case studies and civil servants of other BC communities, linked to partner websites (including BC government websites), shared at academic and practitioner conferences, and also distributed using a quarterly CRC newsletter disseminated to network of over 5000 researchers, practitioners, planners, and government officials across the country. In addition, MC$^3$ researchers shared case study links through blogs and news feeds.

Some case studies contained examples of how climate innovation was driven by community members rather than governing bodies; for example, in Eagle Island (West Vancouver), a local resident spurred climate action by encouraging other community members to engage in local retrofitting projects (Kristensen, 2012). Therefore, in order to inspire similar actions in other communities we developed a diverse library of innovations at multiple scales by multiple actors, and shared this research to the general public through more broad-based communication strategies, i.e., disseminating case study research through popular social media channels (e.g., Facebook and Twitter).

## e-Dialogues

$MC^3$ held two synchronous, real-time online dialogues to share and expand on ideas and discoveries, between the research team, practitioners and local government, and the general public. The project used the e-Dialogue platform, developed in 2001 by the second author, Ann Dale, in conjunction with Isabel Cordua-von Specht and Darren Oxner, for the purposes of exploring the potential the Internet has for engaging diverse groups of people and multiple perspectives in deliberative dialogue on sustainability (Dale, 2005). The e-Dialogue platform is entirely text-based, allowing for accessibility to those with low bandwidth (Dale & Newman, 2006), and thus has the potential to connect people from a variety of different community types, including smaller and rural (which might rely on lower bandwidth connections). This feature was particularly important to the $MC^3$ project as the research examined climate change innovations in communities of varying sizes and urbanness, and many communities in the province are small, single resource-based economies trying to diversify in an expanding global economy.

E-Dialogues can be viewed by the public; however, they are essentially driven by the expert research/practitioner participants (in this case, $MC^3$ researchers and research partners). E-Dialogue audiences are able to provide comments and questions to the dialogue participants; however, this is conducted through a separate forum than the main conversation forum and thus the main conversation and audience feedback are not dynamically interfaced. Therefore, $MC^3$ conducted LiveChats, i.e., public online instant messaging forums, a few weeks following each e-Dialogue as a method of opening a two-way exchange for greater public engagement and adding reflexivity to the research discourse, which is particularly useful for incorporating public values and concerns when engaging in research on science, technology, and innovation (Chilvers, 2013). LiveChats employed a widget (created by CoveritLive) that was embedded on a web page containing information on the respective e-Dialogue. The LiveChat system allowed anyone to type in questions and comments to the researchers without requiring logging into the system (however, questions/comments can be filtered by the LiveChat administrator). The audience comments/questions are posted in the same conversation forum as those of the researchers, thus creating a dynamic two-way dialogue between the research team and the public. In our case, several local government municipal staff participated in the LiveChat, for both educational and policy information purposes.

### Social media

MC$^3$'s use of social media and blogging stems from CRC's previous work in this area. In late 2010, the CRC program began experimenting with social media to evaluate how it could be used for disseminating scientific concepts and research to large, diverse, public audiences. Both authors created a blog specifically to discuss research ideas and concepts, and the program also established a Facebook page, a Twitter account, and YouTube channel, branded as the Humanity, Education and Design (HEAD) Talks (https://www.youtube.com/user/crcresearchRRU). In September 2011, the CRC program began a focused effort in developing their social media platform by creating and releasing HEADTalks videos monthly and posting on Facebook approximately two to three times a week, combined with weekly blogging and tweeting.

We used this prior experience in social media to establish a MC$^3$ blog and employ a CRC Facebook page for sharing ideas and news on climate change and climate innovations. In addition, we produced and published on YouTube a series of five videos that featured animation and interviews with MC$^3$ researchers and research partners and released the series through HEADTalks. The videos covered diverse topics relating to climate change, including the political gridlock resulting from the science "debate" on climate change, the imperative to move to more sustainable development pathways, the importance of and strategies for community engagement, the moral issues surrounding climate change, and the value of peer-to-peer learning exchanges in addressing climate change.

### Virtual conversations

MC$^3$ held three virtual conversations bringing together elected officials from communities across BC, with the research team, to share and facilitate discussion around high-level research outcomes. Officials connected using a bridging service (Blue Jeans Network) that allows people using different methods of communication (i.e., telephone, videoconference, Skype, Google Chat, etc.) to connect in one forum or conference call with full video capability. Because the network supports a variety of methods for communicating, it is convenient, allowing conversation participants to use a method of communicating that they are comfortable with and providing them with alternatives for connecting if one option is not working (i.e., poor Internet signal).

Blue Jeans software was used for several reasons. First, as mentioned above, it offers diverse ways of connecting and is user-friendly, which was important considering that this meeting platform was new to all of the elected officials. People are more likely to engage in a novel activity if obstacles to performing the activity are minimized (Fishbein & Ajzen, 1975), and, through Blue Jeans, the elected officials could engage in the conversation from the convenience of their own office. Second, connecting virtually from their offices allowed them to participate in a safe and confidential environment; confidentiality is important when attempting to facilitate a candid and open conversation among elected officials (Beamer, 2002). Third, considering the discussion was on climate innovation and included participants from across BC, reducing carbon footprints associated with travel through connecting virtually was an appropriate approach to this forum.

## Analysis and Results

### Case studies

Case studies were released in a staggered fashion throughout the months of November 2012–March 2013. This was done to provide more opportunity to advertise and discuss the $MC^3$ project updates, for the purposes of increasing public awareness of the project. This approach is tantamount to how companies and organizations build brand awareness online through advertisement repetition and presence (Yaveroglu & Donthu, 2008). As well, previous experience has shown us the value of regular continuous online engagement, with symmetry and deliberative timed release of complimentary social media.

Case studies traffic was analyzed between 1 November 2012 and 30 June 2013 because the Campbell River study, the first release, was published on the 30th of October and the $MC^3$ project became (mostly) inactive at the end of June following the release of a two-page summary of the Climate Action Agenda. In this period, approximately 22.4% of all $MC^3$ website traffic related to the case studies, including views of the interactive map, case library and news releases for case studies. However, this value does not capture visits to the home page that occurred specifically for the purposes of searching for case studies, and thus a higher proportion of web traffic could have been related to the case studies through indirect traffic.

If the studies all received the same steady and consistent level viewership throughout the release period to the end of June 2013, the proportionate levels of case study views to total web page visits, measured from the respective date of their release, would be expected to be the same for each case. However, strong statistical evidence suggests that these proportions are not similar ($\chi^2$ = 99.176, df = 10, $p < 0.001$). This difference in proportions can be explained by examining how trends in overall visits and viewers of $MC^3$ material affect views of case studies.

Figure 1 shows that case study views for each case increased (overall) from month to month throughout the period, November 2012–June 2013 (inclusive). This indicates that case studies views were not at a constant level, rather consistently built throughout the project. In addition, case studies views did not receive the highest level of viewership right after release, i.e., when they are "new", which would be expected considering this is when a given study was publicized the most. To further investigate why case studies received steady increases in viewership, a regression analysis of overall web traffic was conducted, and this showed the number of visitors to all $MC^3$ content increased steady throughout the case study release period ($R^2$ = 0.854, $F$ = 35.17, $p$ = 0.001). When comparing the number of visitors with case study views through a series of regressions, we found strong evidence for every case (except for Victoria, where there was insufficient data to run a reliable test) that a direct relationship exists between case study views and overall level of web traffic (all $R^2$ values > 0.687, and all $p$-values < 0.05). Consequently, in the case of $MC^3$, the viewership a study received was more tightly associated with the popularity of the project rather than the degree to which study was publicized/advertised, suggesting that traffic can be brought to the case studies effectively through channels other than directly advertising their release (i.e., generating interest in $MC^3$ through blogs, presentations, and learning exchanges; Figure 2).

*e-Dialogues*

Quantitative analysis of e-Dialogues is more difficult to perform as total attendance numbers cannot be accurately ascertained. Most audience members of an e-Dialogue login as a "guest", and a guest login does not register as a unique user in the administration panel. Therefore, albeit the number of people attending an e-Dialogue at a given moment can be measured, whether participants logged in as guests are leaving the conversation while others are entering, also logged in as guests, cannot be determined. Therefore, any estimate of an e-Dialogue is likely lower than the true

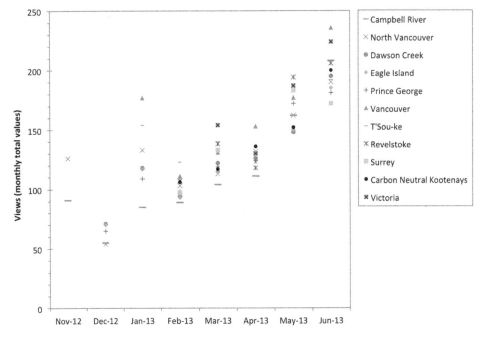

**Figure 2.** Monthly views of $MC^3$ case studies between November 2012 and June 13 (inclusive).

value. However, accurate numbers for LiveChat audiences can be calculated as the number of people who click on the LiveChat widget is recorded through the CoveritLive administration panel.

The highest numbers of audience members recorded for the e-Dialogues at a given moment were 25 and 41 for the first and second e-Dialogue, respectively. Total numbers of attendees for each of LiveChats following the e-Dialogues were 19 and 24, respectively. Keeping in mind e-Dialogue values underestimate the total participation (possibly by a large margin) where LiveChat values definitively express total participation, e-Dialogues can be noted, with confidence, as attracting larger audiences than their LiveChat counterparts.

Attempts were made to collect survey-based data on the effectiveness of using e-Dialogues for mobilizing knowledge and engaging the public. Pre-dialogue and post-dialogue online surveys were prepared for the first e-Dialogue; however, response rates to surveys were too low ($n = 2$ and $n = 1$, respectively) to conduct a reliable analysis on the data. No incentive was offered for completing the survey, which likely contributed to the low response rate (Deutskens, De Ruyter, Wetzels, & Oosterveld, 2004). In addition, it is important to note that response rates for surveys in general are declining as survey requests to the public have increased in recent years (Sheehan, 2001), which affects efforts in conducting social science research in general using online tools.


## Social media

Social media is becoming increasingly more common in the academic sector for connecting the public with research ideas and for gaging the extensiveness to which these ideas have been disseminated (Galloway & Rauh, 2013). The following sections analyze $MC^3$'s use of social media for disseminating research and for connecting to the public with research outcomes and related ideas on climate innovation.


*Facebook.* Facebook data were obtained using Facebook's analytics system. Data used for analysis are associated with Facebook posts created between October 2012 and June 2013, as this period consisted of the highest level of $MC^3$-related Facebook activity (due to video and case study releases). Metrics used for analysis were post views, i.e., number of people who saw the posts and post interactions, i.e., number of times post was shared, reposted or "liking" (a method of actively showing favor toward the post). Analysis of Facebook was conducted using two sets of comparisons. Firstly, $MC^3$-related posts were compared to posts not related to $MC^3$ to determine whether the name/brand of the project particularly attracted audiences. Secondly, posts with climate change related content were compared with posts unrelated to climate change to determine whether the topic of climate change was an audience attractor.

No significant difference between viewership was found when comparing views between Facebook posts that contain $MC^3$-related content and posts that did not

($t = -1.67$, df = 89, $p = 0.099$). However, when reorganizing the data into posts containing climate change related content (i.e., $MC^3$ and other climate change) and those not containing climate change related content, evidence exists suggesting posts not containing climate change content received higher viewership ($t = -2.18$, df = 89, $p = 0.032$). Differences between numbers of interactions did not produce statistically significant results for both corresponding tests ($MC^3$-related content: $t = -1.27$, df = 89, $p = 0.208$; climate change related content: $t = -1.85$, df = 89, $p = 0.068$). Although, it is worth noting that, at an alpha of 0.1, weak statistical evidence exists for a difference in interactions between posts with climate change related content and posts without, but evidence would still not exist for a difference between posts with $MC^3$-related content and that without.

Reorganizing all climate change related posts into one data-set (rather than keeping data-sets organized as $MC^3$-related and not $MC^3$-related) increased $n$ for the climate change data-set (as it combined $MC^3$ and other posts) while decreasing $n$ for the data-set of the remaining posts. As seen in Figure 3, this action widened the gap between mean views of climate change posts and mean views of all other posts. One might conclude from these observations that the topic of climate change received less attention through CRC Facebook than did the other topics on the Facebook page (related to sustainability and sustainable community development). However, further analysis shows that the differences in attention and viewership are more likely related to the type of media featured in the post rather than the topic.

Strong evidence exists to suggest viewership of Facebook posts differ by the type of media featured in the post ($F = 16.13$, df = 90, $p < 0.001$), and, similarly, evidence

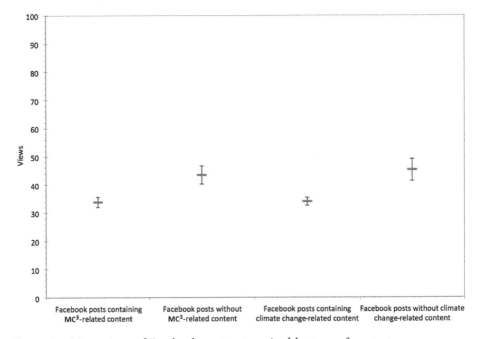

**Figure 3.** Mean views of Facebook posts categorized by type of content.

also exists suggesting post interactions are influenced by media type ($F = 16.04$, df $= 90$, $p < 0.001$). For both views and interactions, Tukey's post hoc analysis identifies posts containing images (i.e., single panel cartoons, mind maps, etc.) as receiving more attention than posts containing embedded videos or links to articles or websites. When separating data into climate change related posts ($MC^3$ and other) and posts not related to climate change and then running analysis of variance tests on each data-set, the media type included in the post was observed to influence viewership for each data-set ($F = 3.68$, df $= 33$, $p = 0.037$ and $F = 17.10$, df $= 56$, $p < 0.001$, respectively).

Our observation that climate change related Facebook content received less viewership than content not related to climate likely was influenced by the fact that the data-set not related to climate change held almost twice as many posts featuring image media. This statement is supported through a series of analyses that separates each of the two data-sets, climate change related posts, and climate change unrelated posts, into categories characterized by the particular type of media featured in the post and then compares the data-sets for each media category. Whether climate change was discussed in the post did not significantly affect viewership, for each media category (link: $t = -0.72$, df $= 56$, $p = 0.472$; video: $t = -1.33$, df $= 20$, $p = 0.199$; image: $t = -2.15$, df $= 9$, $p = 0.060$).

*YouTube.* Figure 4 displays trends in cumulative views (created using seven days, i.e., one week, moving averages) for each of the $MC^3$ videos from when each of the videos

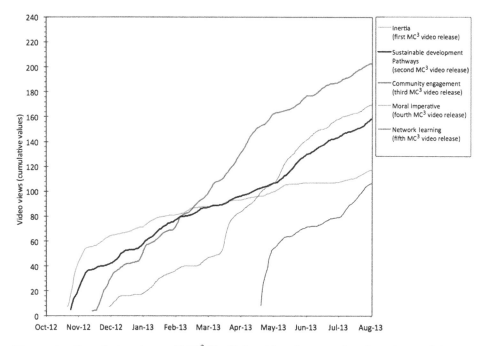

**Figure 4.** Cumulative views of $MC^3$ YouTube videos between October 2012 and August 2013 (inclusive), plotted using seven-day moving averages.

was released to the end of July 2013. Each video consistently received views from the date of their release to July 2013; however, the first video released, *Inertia*, appeared to have a slower rate of increase in views than the other MC$^3$ videos. A possible explanation for this difference in trend could relate to the fact that *Inertia* is the only video that does not have a narrative, meaning it does not feature interviews with MC$^3$ researchers or partners. Sharing and distributing of videos by those who participated in their creation (through interview and/or voice-over) could have contributed to higher rates of accumulated views, which would account for why the one video that does not feature a narrator would have the lowest rate of viewership. This is, however, only a possible explanation that cannot be stated with confidence without conducting further analysis on video sharing through social networks.

Figure 5 displays the release dates and total number of views for each HEADTalks video release over a period ranging from May 2011 (i.e., when the first video was released) to the end of July 2013. Through regression analysis, the general trend dictates that a video released earlier will have received more total views than a video released at a later date ($R^2$ = 0.591, $F$ = 37.58, $p$ < 0.001). This observation is consistent with trends observed in Figure 4 showing that videos continually receive views after they are released.

MC$^3$ videos were compared to the level of viewership predicted for their respective dates of release by the regression model to determine whether they received significantly higher or lower views than the other videos. No significant difference was found between actual and predicted views (paired $t$-test: $t$ = 0.32, $p$ = 0.768).

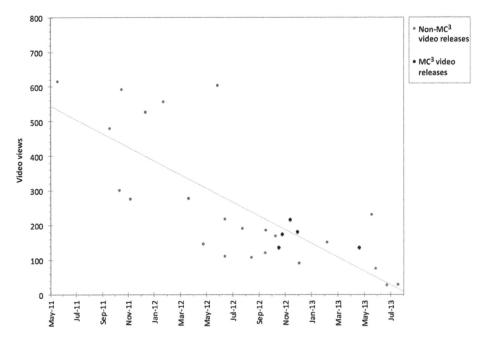

**Figure 5.** Total views of HEADTalks videos (as of August 2013) plotted along the time axis according to date of release.

*Blogging.* Regression analysis produces strong evidence that viewership of the $MC^3$ blog has increased from when the website was made public in September 2011 to when the project became inactive in June 2013 ($R^2 = 0.857$, $F = 120.18$, $p < 0.001$). Further analysis indicates proportions of blog-related views to total $MC^3$ webpages views have changed from month to month ($\chi^2 = 2941.146$, df = 21, $p < 0.001$), and Figure 6 shows this change. A higher proportion of $MC^3$ web traffic was directed toward the blog in the later months of the project (as opposed to the beginning of the project). This observation (in tandem with statistical evidence detailed above) suggests that $MC^3$ blog built audience throughout the course of the project, and blog posts became an increasingly stronger method for communicating with the public.

## Virtual lunchtime conversation

The virtual peer-to-peer learning exchange with elected officials was designed to be an accessible, convenient, and confidential, safe space to encourage participation and candid conversation. Specific participants and communities cannot be identified in this paper due to confidentially reasons; therefore, this analysis is limited to examining participation rate, i.e., number of invitees that participated, which was observed be fairly low, occurring between 14% and 16%. Nevertheless, this exchange was unanimously seen as very positive, all participants asked for more such conversations to occur, novel information was exchanged between the elected

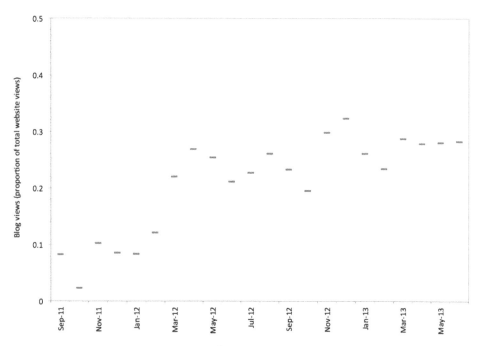

**Figure 6.** Monthly proportions of $MC^3$ website traffic directed to blog plotted over the period of September 2011–June 2013 (inclusive).

officials as well as with the research team, and several key new innovations were explained, which the $MC^3$ research team will now include on its blog.

## Discussion

The insights that emerged from the analysis of $MC^3$'s use of the Internet for research dissemination and knowledge mobilization hold implications for climate change innovation and action; however, they are not limited to this topic. In fact, some of the analyses demonstrated that climate change did not engage audiences differently than other topics in sustainability (see Facebook engagement, discussed in further detail below) and thus these research findings could apply to strategies for disseminating other research outcomes. For example, a major research finding was that developing online presence and "brand" was very effective for ensuring research materials were being used, and this likely is the case for online research projects on a variety of topics, such as community health, social justice, pollution issues, etc. Regardless, due to growing concerns around climate change (Easterling et al., 2000; IPCC, 2014; Stern, 2006), increasing the density and centrality of these issues with climate action would further advance the global dialogue.

Insights emerging from the research can be categorized into four major themes— the effectiveness of building online presence and project awareness; differences between active and passive online audience engagement; the influences media type has on online content engagement; and the relationship (or lack thereof) between convenience and participation in online events. Each theme is discussed in further detail below in terms of the analysis it draws upon and its implications for climate action. The discussion concludes with recommendations on future research.

### Online presence and project awareness

The regressions conducted on the case study data and the visitor traffic to the $MC^3$ website clearly show that the popularity of a website affected the viewership of online research materials. For the purposes of this paper, popularity can be dependent upon many factors, such as attractiveness (aesthetics), ease of use, credibility, trust and reputation of the originators, as well as status. This seems like an intuitive point; however, an interesting observation produced from this analysis is that research case studies received higher levels of views in subsequent months after their release rather than only the new releases incrementally receiving higher views than previous releases, as the website became more popular. This is a different relationship than what is seen with other online materials, such as, for example, public videos. As seen with Figure 4, YouTube videos typically receive the most views right after release and then steady increases of views, thereafter. In contrast, $MC^3$ case studies appeared to continually receive higher numbers of views on a monthly basis, independent of the timing of their release, demonstrating that these materials received more use as the project developed and gained more traction and recognition (i.e., through meetings, dialogues, and presentations), as opposed to being time sensitive to being viewed when publicized.

This increase in the use of $MC^3$ materials is similar to attracting audiences through "brand building." Brand building has been suggested to influence brand loyalty (i.e., "attraction" to a brand), in the online environment (Gommans, Krishnan, & Scheffold, 2001), and, similarly, there is potential for building a name for an online research space, such as the $MC^3$ site. As demonstrated through this research, audience interest in the cases was built for months after the release of the cases, and this was done through building interest in the project and website rather than continually promoting singular case studies, as well as systematic timing and synergy between the channels. This illustrates the power that building an online presence, profile and project awareness has in encouraging practical applications of research, i.e., getting the information to people that can use it, and also shows that audience building efforts have utility long after a research project has ended.

A significant source of "brand attraction" to $MC^3$ was the project blog. The blog received an increasingly larger audience throughout the project, and, furthermore, it progressively became a more significant source of traffic attracting people to the website. Other researchers have made similar observations that the use of social media programs have led to increased readership of their research publications (Bik & Goldstein, 2013), suggesting that activities focused on building audiences and disseminating information do not have to be resource intensive and can consist of actions such as regular blogging and use of Twitter. As aforementioned, building audience and research dissemination is important for climate change research to ensure climate action is informed by the latest research and best practices; therefore, blogging and other such audience building activities are simple actions that might make the difference as to whether research ideas are applied or ignored.

*Online audience engagement*

Through experimentation with online real-time dialogues, it became apparent that e-audiences had a higher proclivity for and were more prone to a spectate-and-learn engagement rather than actively engaging in two-way dialogue. e-Dialogues attracted more participants than their corresponding LiveChats, and where the former consist of dynamic, expertly moderated panel conversations with high appeal to an e-audience, LiveChats are driven by audience questions and participation. In addition, albeit LiveChats attracted audiences of 19 and 24, a minority of the audience (4 and less) actually provided questions and/or comments to the researchers and some of the questions and comments were actually delivered by audience members affiliated with the research project. Because the analysis of this research is limited to the $MC^3$ project, it is difficult to make a broad statement as to why audiences were more attracted to "listening in" (figuratively speaking) to an e-panel conversation over actively engaging in two-way dialogue with the researchers; however, the observation does raise interesting questions regarding public engagement in dialogue around climate change and (more broadly thinking) science. Previous research has identified gaps in establishing dialogue between the science and policy community and the public (Chilvers, 2013; Wynne, 2006), and a separation has been noted between the science community and the general public (Durant, 1999). Therefore, we must consider whether the lack of public dialogue and engagement in science and policy has led to a culture that is more inclined to "learn from the experts" rather than actively engaging in discourse on these topics, which is an essential consideration for global, wide-reaching issues such as climate change that require inclusive, integrative solutions. Findings from this research on MC³ dialogues would be consistent with this theory; however, adequately exploring the theory requires further research.

### Type of media and audience attraction

Similar to Marshall McLuhan's (1964, p. 7) frequently referenced quote, "the medium is the message"; we found that, with modern communication technologies, the media influences the message. When analyzing traffic through Facebook and YouTube, the topic or subject of the content was found to have no influence on people's attraction to the post or video. However, further analysis on Facebook data showed that the type of media presented in a post (i.e., video, article, image) did influence the number of people who viewed and interacted with the post. Therefore, the mention of climate change in a post did not have bearing on the viewership of the posting, whereas the presence of a self-explanatory image, i.e., mind map or single panel cartoon, did.

What these findings suggest is that the type of media can influence the conveyance of a message by attracting more viewers to the content. Aesthetic visuals with clear messages can be effective in attracting audiences and quickly conveying information (Korkmaz, 2009), and, in the case of a social media site that contains a large amount of information (such as a Facebook page), capturing attention and conveying information quickly and efficiently can be advantageous. In particular, mind maps created from outcomes of forums, i.e., the MC³ face-to-face learning exchange, were observed to have especially high levels of online viewership and engagement, which suggests that mind mapping could be an effective method of organizing and packaging ideas for public consumption on complex issues, such as climate change.

### Convenience and participation

Advancements in communications technology have provided new degrees of convenience for meetings and forums designed to bring together geographically dispersed parties and convene large interdisciplinary groups. This created opportunities for easy-access knowledge sharing with less travel required, paving the way for more rapid learning and information exchange on climate innovation and best practices with a lowered carbon footprint. According to behavioral theory, adding ease (in this case, convenience) to the performance of a behavior encourages said behavior (Fishbein & Ajzen, 1975). However, contrary to this notion, the virtual lunchtime conversations had low participation rates in spite of the fact that the meetings were designed to be as convenient as possible for the participants to participate. Low participation in events such as the virtual lunchtime conversation could be both a product of the nature of the "busyness" and attention of elected officials, as well as living in such an information-rich society. Webinars, online forums, and virtual meetings are becoming increasingly simpler and inexpensive to set up, which means that offers and opportunities to attend these events have become increasingly more abundant. Consequently, an event is no longer "convenient" to attend when many such events are available and time is limited, and we are presented with an increased competition for the attention of online audiences (Anderson & de Palma, 2012). Schultz and Vandenbosch (1998, p. 127) describe this situation as "information overload, a state in which the amount of information that merits attention exceeds an individual's ability to process it." Therefore, because of the ease in which knowledge is shared online, strategies and efforts need to ensure that information is "visible" amongst the vast "noise" of online events, websites, and social media channels.

## Conclusion

Our observations reflect the nature of sharing research through the Internet and are not necessarily limited to climate change work; however, although this was not fully captured through this analysis, the topic of climate change itself could have exerted an influence on certain research dissemination and knowledge mobilization efforts. Climate change is a politically divided issue (Nisbet, 2009), and thus interest in forums on climate innovations would be heavily guided by competing interests, peer group ideology (Hoffman, 2011; Markowitz & Shariff, 2012), and of course, social concern. A major objective of MC³'s knowledge mobilization was to increase diffusion and uptake of best practices, and thus these efforts could be considered successful when best practices from a leading community had been shared with a community that is lagging behind in this regard. However, if a community is lagging due to a lack of interest in engaging in climate action, then the community and/or community members might not wish to engage in any online knowledge exchange associated with a climate change project. Therefore, although the current study was useful in identifying how to engage practitioners, local governments, and the wider public that (likely) have preexisting interests in climate action and sustainability in general, future areas of inquiry could explore the nuances around, and difficulties with, disseminating climate innovation research (as compared to other topics that are not as politically divided) and in particular, to groups with deeply held ideological beliefs.

Our observations from analyzing MC³'s Internet-based activities have shown that, in order to effectively communicate and disseminate research findings and share knowledge, researchers should establish a dynamic web presence, consider how the public prefers to engage in learning about research or science, how practitioners (in particular) use the media, and be mindful of the sort of media that attracts particular public audience segments. The primary lesson from this research and analysis is that a research project should not cease activity entirely after the research is completed, and strategies should continue that continually explore creative and engaging methods for building audiences, research dissemination, and knowledge mobilization as there is a considerable lag time in use, we suspect particularly by decision-makers and practitioners. Although this lesson is not limited to climate change research, it is particularly important for climate change innovation and policy, as decision-makers, community leaders, and policy-makers need to be connected with outcomes of research efforts to ensure their decisions and actions are evidence-based and to inform future policy directions and stimulate greater adoption take-up.

## Acknowledgments

We would like to acknowledge the support of the Pacific Institute of Climate Solutions and BC Hydro for funding research operations and activities of the MC$^3$ project.

## References

Anderson, S. P., & De Palma, A. (2012). Competition for attention in the information (overload) age. *The RAND Journal of Economics, 43*(1), 1–25.

Biggar, J., & Middleton, C. (2010). Broadband and network environmentalism: The case of One Million Acts of Green. *Telecommunications Journal of Australia, 60*(1), 9.1–9.17. doi:10.2104/tja10009

Bik, H. M., & Goldstein, M. C. (2013). An introduction to social media for scientists. *PLoS Biology, 11*, e1001535. doi:10.1371/journal.pbio.1001535

Beamer, G. (2002). Elite interviews and state politics research. *State Politics & Policy Quarterly, 2*(1), 86–96. doi:10.1177/153244000200200106

Berkhout, T., & Westerhoff, L. (2013). Local energy systems: Evaluating network effectiveness for transformation in British Columbia, Canada. *Environment and Planning C: Government and Policy, 31*, 841–857. doi:10.1068/c11267

Burch, S., Herbert, Y., & Robinson, J. (2014). Meeting the climate change challenge: A scan of greenhouse gas emissions in BC communities. *Local Environment, 14*, 1–19. doi:10.1080/13549839.2014.902370

Chilvers, J. (2013). Reflexive engagement? Actors, learning, and reflexivity in public dialogue on science and technology. *Science Communication, 35*, 283–310. doi:10.1177/1075547012454598

Dale, A. (2001). *At the edge: Sustainable development in the 21st century.* Vancouver: UBC Press.

Dale, A. (2005). A perspective on the evolution of e-dialogues concerning interdisciplinary research on sustainable development in Canada. *Ecology & Society, 10*(1), 1–9.

Dale, A., & Newman, L. (2006). E-dialogues: A role in interactive sustainable development? *The Integrated Assessment Journal, Bridging Sciences and Policy, 6*(4), 131–141.

Dale, A., Robinson, J., Herbert, Y., & Shaw, A. (2013). *Climate change adaptation and mitigation: An action agenda for BC decision-makers. Developed through the Meeting the Climate Change Challenge research partnership.* Royal Roads University, University of British Columbia, and Simon Fraser University. Retrieved April 22, 2013, from http://mc3.royalroads.ca/sites/default/files/webfiles/MC3%20Climate%20Action%20Agenda.pdf

Deutskens, E., De Ruyter, K., Wetzels, M., & Oosterveld, P. (2004). Response rate and response quality of Internet-based surveys: An experimental study. *Marketing Letters, 15*(1), 21–36. doi:10.1023/B:MARK.0000021968.86465.00

Durant, J. (1999). Participatory technology assessment and the democratic model of the public understanding of science. *Science and Public Policy, 26*, 313–319. doi:10.3152/147154399781782329

Easterling, D. R., Meehl, G. A., Parmesan, C., Changnon, S. A., Karl, T. R., & Mearns, L. O. (2000). Climate extremes: Observations, modeling, and impacts. *Science, 289*, 2068–2074. doi:10.1126/science.289.5487.2068

Few, R., Brown, K., & Tompkins, E. L. (2007). Public participation and climate change adaptation: Avoiding the illusion of inclusion. *Climate Policy, 7*(1), 46–59. doi:10.1080/14693062.2007.9685637

Fishbein, M., & Ajzen, I. (1975). *Belief, attitude, intention, and behavior: An introduction to theory and research.* Reading, MA: Addison-Wesley.

Galloway, L. M., & Rauh, A. E. (2013). Social media and citation metrics. *Libraries' and Librarians' Publications.* Paper 107. Retrieved from http://surface.syr.edu/sul/107

Girvetz, E., Zganjar, C., Raber, G., Maurer, E., Kareiva, P., & Lawler, J. (2009). Applied climate-change analysis: The climate wizard tool. *Plos One, 4*(12), e8320. doi:10.1371/journal.pone.0008320

Gommans, M., Krishnan, K. S., & Scheffold, K. B. (2001). From brand loyalty to e-loyalty: A conceptual framework. *Journal of Economic and Social Research, 3*(1), 43–58.

Hoffman, A. J. (2011). The growing climate divide. *Nature Climate Change, 1*, 195–196. doi:10.1038/nclimate1144

Intergovernmental Panel on Climate Change (IPCC). (2007). *IPCC Fourth Assessment Report (AR4): Climate Change 2007. Contribution of Working Groups I, II and III to the Fourth Assessment Report of the Intergovernmental Panel on Climate Change.* Cambridge: Cambridge University Press.

Korkmaz, O. (2009). Primary perceptual field in visual materials. *The Social Sciences, 4*, 525–533.

Kristensen, F. (2012). *Eagle Island. Meeting the climate change challenge (MC3). Online case study.* Vancouver: Royal Roads University, University of British Columbia, and Simon Fraser University. Retrieved November 22, 2012, from http://www.mc-3.ca/eagle-island

Kurtz, C. F., & Snowden, D. J. (2003). The new dynamics of strategy: Sense making in a complex complicated world. *IBM Systems Journal, 42*, 462–483. doi:10.1147/sj.423.0462

Markowitz, E. M., & Shariff, A. F. (2012). Climate change and moral judgement. *Nature Climate Change, 2*, 243–247. doi:10.1038/nclimate1378

McLuhan, M. (1964). *Understanding media: The extensions of man.* New York, NY: McGraw Hill.

Newell, R., & King, L. (2012). *Prince George. Meeting the climate change challenge (MC3). Online case study.* Vancouver: Royal Roads University, University of British Columbia, and Simon Fraser University. Retrieved November 29, 2012, from http://www.mc-3.ca/prince-george

Newman, W. L. (2003). *Social research methods: Qualitative and quantitative approaches.* Boston, MA: Allyn & Bacon.

Nisbet, M. C. (2009). Communicating climate change: Why frames matter for public engagement. *Environment: Science and Policy for Sustainable Development, 51*(2), 12–23. doi:10.3200/ENVT.51.2.12-23

Schultz, U., & Vandenbosch, B. (1998). Information overload in a groupware environment: Now you see it, now you don't. *Journal of Organizational Computing and Electronic Commerce, 8*(2), 127–148. doi:10.1207/s15327744joce0802_3

Sheehan, K. B. (2001). E-mail survey response rates: A review. *Journal of Computer-Mediated Communication, 6*(2), 0. doi:10.1111/j.1083-6101.2001.tb00117.x

Smith, P. J., & Smythe, E. (1999). Globalization, citizenship and technology: The MAI meets the Internet. *Canadian Foreign Policy Journal, 7*(2), 83–105. doi:10.1080/11926422.1999.9673213

Stake, R. (1995). *The art of case study research.* Thousand Oaks, CA: Sage.

Stern, N. (2006). *Review on the economics of climate change.* London: HM Treasury.

Wenkel, K., Berg, M., Mirschel, W., Wieland, R., Nendel, C., & Köstner, B. (2013). LandCaRe DSS – An interactive decision support system for climate change impact assessment and the analysis of potential agricultural land use adaptation strategies. *Journal of Environmental Management, 127*, S168–S183. doi:10.1016/j.jenvman.2013.02.051

Wynne, B. (2006). Public engagement as a means of restoring public trust in science-hitting the notes, but missing the music? *Public Health Genomics, 9*(3), 211–220.

Yaveroglu, I., & Donthu, N. (2008). Advertising repetition and placement issues in on-line environments. *Journal of Advertising, 37*(2), 31–43. doi:10.2753/JOA0091-3367370203

Yin, R. (2003). *Case study research: Design and methods* (3rd ed.). Thousand Oaks, CA: Sage.

# Exploring the Use of Online Platforms for Climate Change Policy and Public Engagement by NGOs in Latin America

Bruno Takahashi, Guy Edwards, J. Timmons Roberts & Ran Duan

*The engagement of environmental non-governmental groups in collaborative communication efforts and decision-making about climate change remains a significant challenge. In Latin America, the website Intercambio Climático was set up to attempt to play a breakthrough role in the region's discussion of climate change. This case study focuses on the development, accomplishments, and challenges of this unique online communication initiative among non-governmental organizations working in climate change issues in Latin America with a US-based partner, as a vehicle to achieve their goals. We used secondary data and in-depth interviews to examine this platform's role in disseminating information and influencing decision-making. The results suggest that generally the participants perceive the website positively, but there are also problematic aspects of the collaboration that are not fully recognized by the members, which prevents a more functional and effective communication strategy.*

Online platforms have emerged as important alternative forums that can successfully overcome individual and societal level barriers and provide opportunities for the engagement of a variety of voices (Cox, 2012). However, there is limited understanding about the ways in which such platforms can be successfully used by non-governmental organizations (NGOs) and whether they can be effective in regions of the world where strong political, organizational, and technological barriers could prevent them from succeeding (Carvalho & Peterson, 2012).

In the case of Latin America, the website *Intercambio Climático* was set up by NGOs to attempt to play a breakthrough role in the region's discussions about climate change (*Intercambio Climático*, n.d.). It was assumed that establishing an online platform may have a wider reach, beyond just the small group of climate experts typically informed about crucial debates. This aim of a wider reach has to be seen in the context of increasing but still relatively low levels of Internet penetration in most of the developing world (compared to the developed world).[1]

However, even in the developed world the availability of these new virtual platforms does not automatically mean that they will yield better communication and policy outcomes or less conflicting collaborative processes. Despite the extensive embrace of these new technologies by many environmental and civil society groups, there is still limited evidence of the value of such approaches. Pickerill (2001) suggests that the use of the Internet as a networking tool has allowed environmental organizations to mobilize international participation, share solutions and ideas, and draw strength from each other. However, such evidence is scarce for developing nations, as well as multinational efforts, which are key components of climate change deliberations.

This study is situated at the crossroads of international climate policy-making, NGO activism, online communication, and new information technologies. Rapid developments in online media have created a unique communicative environment for various stakeholders involved in climate change discourse and action. This has required a reconsideration of the techniques and strategies for public participation and engagement of NGOs and individuals. We provide a first empirical look at the use of an online platform dealing with climate change in Latin America. The focus is on the challenges faced by NGOs in the development of the online tool, and on the effectiveness of such a platform in disseminating information used by member organizations in their advocacy efforts at national and international arenas.

## The Internet and Environmental Activism

Climate change has become one of the most salient global issues in recent human history (McCright & Dunlap, 2003). The complexity of the problem has prompted highly contentious discussions not only of climate change as a scientific issue but also of political and communication strategies used to foster public engagement with climate change (Boykoff, 2009; McCright & Dunlap, 2003). In this context, NGOs play an important but equally contested role. Nevertheless, some scholars argue that the role of NGOs and other civil society groups has been restricted due to their limited access to the decision-making arena (see Carvalho & Peterson, 2012; Endres, Sprain, & Peterson, 2009).

The role of NGOs in the decision-making process is vital if governments are to implement wide-reaching measures to both mitigate greenhouse gases and adapt to the impacts of climate change, especially since other actors, especially businesses and governments who need tax revenues and create economic growth and employment, are unlikely to take on the difficult work of decarbonizing their economies without external pressure. With this in mind, the role of NGOs in climate change governance is important from a perspective of accountability (Newell, 2008). Newell (2008) explains that this oversight function by NGOs and civil society groups should not replace public democratic oversight of power exercised at the international level, but that it should function to supplement the means to achieve political goals in the face of state inaction. In addition, NGOs face increasing difficulties in accessing closed doors meetings which decreases their ability to promote their positions. This pushes them to "use more indirect strategies to keep abreast of negotiations they can no longer observe" (Betsill & Corell, 2001, p. 70).

Environmental groups and advocates therefore seek ways to influence policy-making and to engage different publics in climate change efforts through social movements, political campaigning, and voluntary behavior changes (Newell, 2008). One recent example is Bill McKibben's 350.org campaign. Feldpausch-Parker, Parker, and Peterson (2012) examined the orchestration of thousands of events worldwide and the use of a unified social identity and observe that: "[350.org] fostered a strong sense of global unification that may be especially important given the relative failure of any officially sanctioned forward movement at the Copenhagen summit" (p. 211).

Many of these efforts in climate change advocacy campaigns rely on online technologies to reach their intended audiences. Online media can disseminate information to large audiences very quickly; breach geographical barriers; reduce costs and censorship; facilitate contact with potential donors and volunteers; network between chapters; help in fundraising efforts; and, to a certain extent, reduce personal accountability (Brunsting & Postmes, 2002; Taylor, Kent, & White, 2001). In other words, collaborative online efforts can bring people together that would otherwise not be able to meet due to geographical, physical, or financial limitations (Kurniawan & Rye, 2014). However, there are also valid concerns about the ability of such efforts to overcome ideological, psychological, and power-based barriers among participants. In this respect, some research suggests that online collective efforts can be explained from a social identity theory perspective (Postmes & Brunsting, 2002). In this study we contend that *Intercambio Climático*, the website of the Latin American Platform on Climate (LAPC), has attempted to create a unifying identity and image of a cohesive Latin American coalition on climate change, a way to create an epistemic community (Gough & Shackley, 2001), which until its creation had been missing in the region.

The Internet presents a paradox since it does not allow physical social interactions, but at the same time can create and reinforce social unity. Some recent research shows that the networking activities of online social movements primarily relate to the formation of social and collective identities (Ackland & O'Neil, 2011; Xie, 2011). Brunsting and Postmes (2002) and Postmes and Brunsting (2002) explain that a social identity can be created regardless of the physical presence or proximity of other group members. Moreover, online efforts can mobilize and stimulate collective action if the social identities are made salient and are supported with strategic conditions that empower such identities. Brunsting and Postmes (2002) argue that "The Internet changes the nature of collective action, but contrary to popular belief, the Internet would appear to be especially suited to persuasive collective action rather than confrontational action" (p. 550).

However, some evidence also suggests that online efforts are more effective in facilitating organizational activities of social movements rather than the symbolic activities discussed above (Wall, 2007). Many online platforms are especially set up by activists to create international networks (Lester & Hutchins, 2009). Postmes and Brunsting (2002) report that online activism plays an important role in the recruitment of more peripheral, non-activists to the discussion. However, there is also some evidence of similarities between online and off-line activism. For example, Ackland and O'Neil (2011) found that online networks reflect fragmentation that is similar to off-line social groups. In addition, Bennett (2012) discussed the limitations of online activities in terms of the lack of availability of a visible "face" or central leaders, and the loose connections provided by these networks. These competing conclusions can be explained based on the issue, the characteristics of the groups, and the sociopolitical context, among others, which suggests a need to further explore the assumptions of when and how the web fosters environmental engagement.

Online and social media serve not only as social networks, but also as news source. As Lester and Hutchins (2009) argue, the Internet has the potential for "independent information distribution devoid of the mediating effect of news journalists" (p. 579), and it may therefore be more useful for building distinctive online social movements. However, Lester and Hutchins (2009) argued that the use of new media in environmental activism has not created new ways of communicating or media power, and that the traditional media are more resilient than originally thought.

More importantly, environmental activist groups occasionally benefit greatly from the Internet as it allows two-way communication, promoting dialog as the groups set about engaging the public on their various issues (Kent, Taylor, & White, 2003; Taylor et al., 2001). Some observers argue that some groups have been able to reduce the slope of the political playing field and to resist containment by their opponents (Pickerill, 2001). Pickerill (2001, p. 367) summarizes the main processes of mobilization through the use of the Internet by environmental organizations in the UK: "using the Internet as a gateway to activism; using it to raise the problem of group campaigns; stimulating local activism; mobilizing online activism; and attracting participants to existing protests" (2001, p. 367)

Despite these opportunities, there are also important limitations in this new two-way communication process. Waters, Burnett, Lamm, and Lucas (2009) show that most environmental groups using social media had failed to develop a cohesive policy or plan. Most groups did not take full advantage of the interactive nature of social networking, and only posted links to external information on other websites. Bortree and Seltzer (2009) report similar results, which indicate that environmental advocacy groups may need to be more active in terms of what they post and communicate (i.e. text, photos, videos, and information about events) in order to nurture healthy relationships and discussions with their members and followers.

Maintaining and updating online communication is important, as it can have an influence on policy and decision-makers. For example, there is some evidence that the Internet is one of the main sources of information for some policy-makers about climate change. A study by Takahashi and Meisner (2013) showed that legislators in Peru rely heavily on the web to seek information and learn about climate change and to draft policy proposals, mostly because of the limited availability of alternative sources of science and policy information from credible sources (e.g. researchers, NGOs, government institutions, etc.).

In summary, and as noted by Bach and Stark (2004), new forms of communication technology may promote an expanded and vibrant civil society—but not radically transform it—because new media for communication enable new forms of representation in a virtual public sphere. As a result, the use of online media and communication by NGOs and environmental groups can play an important role in public engagement and policy decision-making about climate change and other environmental issues. However, there is limited evidence about the ways in which the processes of identity formation, mobilization and advocacy, information dissemination, and social influence, among others, take place in places where technological, cultural, and political factors differ from those in the USA or Europe.

## Theoretical Considerations

Some of the research reviewed above suggests that NGOs can, under certain circumstances, influence global environmental politics (Betsill & Corell, 2001; Betzold, 2010; Humphreys, 2004). However, there is no clarity about how and under what conditions this influence happens. First, there are limitations in terms of the definition of "influence" as well as a lack of evidence that shows a causal mechanism to measure and assess such an influence (Betsill & Corell, 2001). This influence can be examined, for example, by looking at policy documents, or agreements, and the extent to which claims or requests by NGOs are present. Betsill and Corell (2001) have examined issues around influence in the context of international environmental negotiations. These researchers argue that NGOs influence international environmental negotiations when they intentionally and successfully transmit information to negotiators that alters both the negotiating process and outcome from what would have occurred otherwise. Corell and Betsill (2001) comparatively studied two environmental negotiation issues, desertification and climate change to examine how influence happens. Their findings suggest that the nature, history, and framing of the issue under negotiation are all important, as well as the political opportunity structure and the NGO profile. However, the analysis of framing focused specifically on the dissemination of information via online activities, and has not been properly incorporated in the analysis of environmental negotiations.

The notion of NGO influence has therefore two dimensions: (1) the intentional transmission of information by NGOs; and (2) alterations in behavior in response to that information. Regardless of the power inequality, "influence" is defined as the relationships between actors. In this study, we focus mostly on the first dimension, which is the intentional transmission of information by NGOs, focusing on the role of online information as a way to increase information availability, which is something not entirely developed by Betsill and Corell (2001). This change is justified because online platforms have arisen as important communicative forums that build up transnational cooperation and presumably increase NGO influence. Betsill and Corell (2001, p. 79) argued that researchers should focus on these questions: What did NGOs do to transmit information to decision-makers? What opportunities did NGOs have to transmit information? What sources of leverage did NGOs use to transmit information? The second dimension, related to behaviors, is only briefly discussed in this paper, as the data collected for this study does not allow for a more in-depth examination.

Our focus is on the actions and initiatives undertaken via the online platform *Intercambio Climático*. We explore the strategic communication goals and processes. At a secondary level, we discuss some of the impacts of these actions, as perceived by some participants of the initiative. With that in mind, the following research questions, based on the discussion presented above, guide this study: (1) In what ways does *Intercambio Climático* serve as a vehicle for NGO groups in Latin America to communicate and collaborate about climate change policy and decision-making?, (2) What are the barriers to effective public engagement of NGOs in this online effort?, and (3) In what ways do NGOs utilize this online media platform to learn about climate change issues, and to advance their policy agendas?

## The Case Study: *Intercambio Climático*

The above discussion points toward some areas of research in online media communication, policy-making, and public engagement that have not been explored in much detail in the case of climate change, and especially in developing nations. Although the debate on climate change in Latin America has increased rapidly over the last few years, the information in the public domain and conversations between decision-makers and civil society is insufficient and relatively new. This is despite the fact that Latin American citizens on balance are more concerned about climate change with little climate skepticism compared to their northern neighbors (PEW, 2013). Within this context, this case study explores the role of *Intercambio Climático*,[2] Latin America's first multilingual website focusing on climate change, in order to better understand the potential for online communication in influencing policy change and public engagement.

Latin America is particularly vulnerable to climate change impacts, including the potential collapse of the Caribbean coral biome, intensification of weather patterns and storms, rising sea levels, increased flooding and droughts, warming of high Andean ecosystems, increased exposure to tropical diseases, and risk of dieback of the

Amazon rainforest ecosystem (*El Cambio Climático No Tiene Fronteras*, 2008; ¿*El Fin de las Cumbres Nevadas? Glaciares y Cambio Climático en la Comunidad Andina*, 2007) Latin America's vast natural resources are simultaneously a major asset and an Achilles' heel. The region is extremely vulnerable to ecosystem transformations resulting from climate change, as its economies are dependent on raw materials and natural resources (*El Cambio Climático No Tiene Fronteras*, 2008).

Latin American countries do not speak with one voice at the UN climate change negotiations (Edwards, 2013). The region has over half a dozen groups. For example, Brazil is member of BASIC that is a group of four countries Brazil, South Africa, India, and China, whereas Mexico is a member of the Environmental Integrity Group with South Korea and Switzerland. Venezuela, Bolivia, and Ecuador negotiate as part of the Bolivarian Alliance for the Peoples of Our America. Colombia, Peru, Costa Rica, and Chile make up the Independent Association of Latin America and the Caribbean. However, many Latin American countries have played diverse and crucial roles at the UN climate negotiations despite this fragmentation (Edwards & Roberts, 2015).

The issue of climate change has steadily increased in salience in nearly all Latin American countries over the last decade. As Ryan (2012) points out, across the region important steps have been taken in policy-making on climate change and in the creation of specific institutions to attempt to tackle the issue. However, there is a major deficit in the implementation of these governance measures and climate policies are only weakly integrated with other policies (Takahashi & Meisner, 2013). Ryan (2012) suggests the existence of institutions and actors within the state that focus on climate issues and with technical capacity and access to international resources (e.g. funding), are essential to sustain activities on climate change overtime and across different administrations. He argues there is also a widespread recognition across governments and NGOs of the importance of citizen participation on climate change.

In 2008, the private Avina Foundation (founded by the Swiss billionaire Stephen Schmidheiny) approached several influential environmental NGOs in South America with the goal of creating a network to develop climate change initiatives. A total of 17 NGOs[3] represent the following countries: Argentina, Bolivia, Brazil, Chile, Colombia, Ecuador, Paraguay, Peru, and Uruguay. The Avina Foundation provided some seed money to start the initiative, which included meetings among its members to design the structure of the effort and its strategic plan. The network was officially launched in January 2009 as the *Plataforma Climática Latinoamericana* (LAPC). The main objective of the LAPC is to contribute to "ensuring that addressing climate change and its effects will be a top-priority for political, environmental, social and economic decision-making, at multiple levels, in both the public and private sector".[4] It aims to provide an open space for dialog concerning the international debate on climate change. The LAPC is principally a voluntary effort by most of its members, and communication activities are for the most part limited to the individual efforts of core staff members of its member groups via their existing contacts and networks.

In 2010, researchers at Brown University's Center for Environmental Studies (CES) approached the *Fundación Futuro Latinoamericano* (FFLA), then the Executive Secretariat of the LAPC, with a proposal to create a new website for the Platform. Discussions between CES and FFLA focused on how the LAPC could benefit from such a tool, which could offer opportunities for its members to share information and knowledge with each other and with other NGOs, governments, think tanks, international organizations, and the general public. The idea was that such a website could provide an opportunity to open a space for dialog on climate change across countries and actors, and to raise the profile of the region in international discussions on climate change. A new website could also complement the objectives of the LAPC by assisting in improving its visibility and ability to promote and facilitate dialog with climate change policy-makers. Following this exchange, a team from CES cofounded *Intercambio Climático* in partnership with the LAPC. The online website was launched at the United Nations Climate Change Conference in Cancún, Mexico, in December 2010.

With the number of Internet users in Latin America rapidly increasing and the rise of social media globally and its popularity in Latin America, the LAPC could use this transformation to meet some of its objectives. However, despite the fact that almost 50% of people in the region now access the web, the digital divide still disproportionately excludes the poorer and most vulnerable populations from engaging in these online conversations. Despite this information and technological gap, various surveys illustrate that Latin American citizens are concerned by climate change. A survey carried out by Gallup in 100 countries in 2011 indicated that Mexico, Colombia, Venezuela, and Ecuador are amongst the countries most concerned by global warming. Another poll conducted by Nielsen among 25,000 Internet users in 51 countries concluded that in Latin America concern about climate change was expressed by 90% of those consulted while the global average was 69% (Martins, 2011).

The design of *Intercambio Climático* was inspired by another online platform called *Global Dashboard*, a leading global affairs blog. The website was originally intended to be the clearinghouse of climate change and Latin America. Visitors would be able to locate materials on a wide range of themes related to climate change (e.g. mitigation, adaptation, REDD+ [Reducing Emissions from Deforestation and Forest Degradation], etc.) and the region. This flexible approach from the outset was an attempt to encourage greater participation by LAPC members to offer a wide range of possible topics and formats for them to write on and to cater for the diverse interests of the online audience. Further, the website is an attempt to provide the LAPC with an attractive and user-friendly online space to promote the identity, objectives, activities, and research outputs of their network. Providing a space for LAPC members to publish their work was also an attempt to encourage greater output of articles on climate change. It was hoped that having an attractive space available would be an incentive to contribute regularly to the website. It also serves to connect the LAPC and its members with an online audience especially policy-makers, researchers, and the media. The website did provide space for visitors to comment which received modest interest but this space has subsequently been closed due to a security breach which compromised the website.

## Methods

This study follows a case study approach (Creswell, 2012; Yin, 2009). We focus on *Intercambio Climático* as a unique online media space in Latin America to investigate its weaknesses and strengths, as well as the barriers to widespread participation and adoption of the medium. We recognize the limitations in terms of generalizability of single case studies, but we also want to emphasize the importance of researching unique initiatives such as *Intercambio Climático* as a first step in researching a largely untapped area of study. We therefore do not make generalizations of the results of this study, but we do point out future areas of research.

The study uses mixed methods to answer our research questions. We follow the approach discussed by Ackland and O'Neil (2011), who examined obtrusive (e.g. interviews) and unobtrusive (e.g. web content analysis) methods of online social movement research. First, we utilized secondary data to provide a descriptive account of the development of the website, including its objectives, and the exchange of information and agreements among the various organizations involved. Second, we looked at the web analytics that were available to us to provide an overview of the reach of the website and the origins of its contents. Finally, we conducted in-depth interviews with a sample of individuals from the LAPC and other actors who contribute to the website.

The interviewees were selected from the list of all active participants of the LAPC. This list included individuals from various organizations in the following countries: Argentina, Bolivia, Brazil, Chile, Colombia, Ecuador, Mexico, Peru, and Uruguay. We identified 12 individuals who are part of the LAPC, and 5 individuals identified as content contributors or "invited authors" to the website. We then contacted them by email and invited them to participate in the study. After an additional reminder, seven agreed to be interviewed. Four of the interviewees are current members of the LAPC, and three are invited authors to *Intercambio Climático*. Interviews were conducted by one of the authors via telephone and Skype during the month of October 2013. The Appendix lists the position and organization of each interviewee, as well as a code that is used to attribute the quotes presented in the results and discussion sections.

The interviews lasted between 30 and 60 minutes, and followed a semi-structured format. Questions focused on the interviewees' participation in the LAPC and *Intercambio Climático*, their perspectives on the benefits, limitations, and challenges of online collaboration. The interviews were audio recorded and then transcribed for analysis in Spanish. The analysis was conducted on the Spanish transcripts to avoid losing meanings in the translation process. Quotes that were selected for the manuscripts were later translated by one of the authors who is a native speaker of Spanish. The analysis followed an iterative coding procedure, where codes were developed inductively by constantly reading the transcripts and highlighting and selecting common themes among them.

Based on the previous theoretical discussions, the data from the interviews were analyzed by looking at several factors of importance for the way the use of the Internet was organized within the Environmental Non-Governmental Organizations (ENGOs), including: financial support, degree of specialization, infrastructure, and technology perception, among others (Kurniawan & Rye, 2014).

## Results

We begin with two broad observations by LAPC members on the role of LAPC in climate change policies:

> I think the LAPC gave us the opportunity to get together in a room and communicate these experiences or that knowledge for the benefit of Latin America ... and start building a slightly more regional vision. (SG)
>
> I think the LAPC has established itself as a leader inside and outside the continent. There is no regional non-governmental organization that carries on the voice of NGOs; understood the "non-governmental" as ordinary citizens, scientists, businessmen. (AM)

These quotes exemplify the overall feeling about the LAPC among those interviewed. The LAPC is perceived as a place where individuals can share information, experience different perspectives, compare processes, receive feedback, and follow trends. Moreover, some respondents also believe that the platform has been successful in partly achieving one of its main objectives, which is to influence the debate among policy and decision-makers.

This success has been built through the achievement of various objectives throughout the last couple of years, but three specific accomplishments of the LAPC that served as key topics of diffusion via *Intercambio Climático* are particularly relevant. First, funded by the Oak Foundation and the Climate and Development Knowledge Network, the LAPC commissioned a series of studies on national public policies about climate change in all member countries of the LAPC. These studies and subsequent reports focused primarily on climate change and forestry and agriculture. The reports are the first of their kind by a civil society network in Latin America and attempt to identify the existing government policies that address climate change and development in each of the chosen countries.[5]

At the UN Climate Change Conference in Warsaw, Poland, in November 2013, the LAPC co-organized an event attended by over 130 people where the results of the reports were shared.[6] In addition, members of the LAPC and those contracted to assist with the writing of the reports also held meetings with government officials in their respective countries to discuss the scope and findings of the reports, which were then published on the *Intercambio Climático* website for wider distribution:

> There are 10 countries studied, with a uniform methodology, agreed, consensual, and the possibility ... well, it had the opportunity to compare climate policies of the 10 countries studied. That for example is somewhat unusual, unprecedented, to my

knowledge there is no such a thing, or an institution able to do it with that value. (AM)

The second achievement corresponds to the decision to invite a number of leading experts and specialists from Latin America on climate change to contribute to the website. In some cases the invited authors can be regarded as leading thinkers within governments and NGOs (e.g. former president of Chile, Ricardo Lagos). Former President Lagos wrote about the potential for social media to support efforts to confront climate change in Latin America and also the importance of Latin American countries' climate diplomacy to achieve success at the UN climate change negotiations in Durban, 2011. Other guest authors represent a mixture of United Nations Framework Convention on Climate Change (UNFCCC) negotiators from Latin American countries, journalists and civil society representatives.

A focus of the contributions of many of these guest authors was the UN climate change negotiations and particular aspects such as climate finance. In addition, these guest authors presented important country perspectives from Latin American countries that have a weak or nonexistent presence in the LAPC, such as Costa Rica and Mexico. As discussed in more detail later, the caliber and area of expertise especially of former president Lagos and negotiators to the UNFCCC helped inadvertently to create a type of virtual think tank.

The third achievement relates to the selection of Peru as the host nation for the meeting of the Twentieth Conference of the Parties (COP20) in 2014. In June 2013, researchers at the CES in conjunction with the LAPC drafted a declaration in support of the candidacy of Peru (over Venezuela) to host the COP20 in 2014. Peru and Venezuela had both put forward their candidacies to host COP20. In the interests of the negotiations, Peru is considered the stronger candidate and Venezuela's role has been more controversial. In 2008, Peru became the first developing country to offer a voluntary emissions reduction target, which was extremely well received.[7] The declaration states that Peru presents a very promising candidacy to host COP20 in 2014. Peru has been a progressive voice in the formal and informal discussions on climate change and has demonstrated a willingness to listen and offer counsel in the international negotiations on climate change.[8] The declaration was signed by over 60 NGOs from across Latin America, including the LAPC. The declaration was published on *Intercambio Climático* and was widely distributed across various networks.

According to the interviewees, the declaration helped influence the final selection of Peru over Venezuela at a meeting of the Latin America and Caribbean Group (GRULAC) during the June 2013 meeting of the UNFCCC in Bonn, Germany. Although we do not know exactly how the declaration influenced the final decision, given the private nature of the GRULAC meeting, we were informed by a Latin American country negotiator that members of the GRULAC meeting were aware of the declaration prior to the final decision and that it supported the consensus in the room in favor of Peru. We can therefore assume that the publication of the declaration on the *Intercambio Climático* website at the very least made the declaration more readily available to negotiators from the region, some of whom are colleagues or know the guest authors featured on the website. This may reflect the reason behind sponsoring organizations' authorization to publish the declaration online, since they felt that given the Internet's reach and unique position of the website it might stand a better chance of being picked up by GRULAC negotiators.

One respondent observed that:

> I know several activists in Peru since before COP16 in Mexico, and I think it was a movement that has been increasingly integrated, and this time I think the articulation of civil society in Peru was the trigger for other networks around this network of Peru say: "Hey, well, see, the network will be in Peru, there is activism in Peru, and it happens to be a good scenario for Latin America, we must support it." (SG)

Part of the success described above is due to the involvement in the LAPC of influential members at the national level. In the case of Peru, the former president of the Peruvian Society of Environmental Law (LAPC member organization), Manuel Pulgar Vidal, is now the Peruvian minister of the environment and currently the president of COP20. In addition, some of the success is also attributed to the visibility achieved by the LAPC via the website *Intercambio Climático*:

> the blog has really been a very useful tool; I have been following the international negotiations on more than one occasion, and people from other places have made reference to the [website]. So I think it is a tool that has worked quite well, which has been positioned pretty well, even internationally. (EM)
>
> [T]he platform as part of this digital revolution is very important, it is very useful, necessary, and I think we're in a time when we need more reliable media, that is, even in this digital era one finds information of quite low quality also in the media, then you always need a place where they can see first-hand information but of high quality. (SG)

Since the launch of the website in December 2010 up until the writing of this article in September 2013, roughly 77,000 people have visited the site a total of 97,698 times with an average visit of 00:01:55 minutes. These visits account for 167,472 page views. Viewed on a monthly basis the number of people visiting the site has generally fluctuated between 3000 and 5000 visits, which compared to other websites specializing on climate change is significantly less.[9] However, given the regional focus compared to the global nature of the other websites, and the relatively smaller number of climate experts in the region, these numbers may not be insignificant. While policy-makers and climate experts in the region and beyond may visit the site, these numbers temper any claims of mass public impact. In addition, the data do not provide any indication of the characteristics of the visitors, making any general-izations of the reach or impact impractical.

Some of the interviews highlighted the importance of the website in creating a sense of community and shared identity based on a Latin American perspective on climate change in order to create a successful collaborative process. In this respect, some respondents recognized that online media could play this function:

> [Online media] are significant and should not be missed, plus they are so cheap, so inexpensive, so interactive, allowing to not only connect you, but to form a community with your audience, but I also worry about the others that are not [connected]. (AM)

The role of the CES team in this process was crucial for the original conception of the website idea and there was an overall recognition by the interviewees about the commitment and quality of that team's work in developing, designing, and managing the website. An external web designer built the site under the direction of the CES and LAPC. The LAPC came up with the website's name. Even though no member of the Brown team comes from Latin America (although one of them was based in Quito, Ecuador, from 2010 to 2013), there was an overall acceptance of their role and contribution. This could be seen as a success in removing a perceived barrier, but there is not clarity from the interviewees about the extent to which the partnership had an effect on the limited identification with a unified identity:

> (he) has achieved things that are very difficult to achieve in Latin America. One of them, despite being an Anglo-Saxon, to feel that he is one of us ... he was not just another consultant, he was not just another international aid worker, there are even e-mails from the platform on which they say "(he) is a friend"... that made possible the success of the project, created the conditions, provided the playing field, provided a suitable ground for the emergence of this blog and its success. (AM)

### Content production and information use

The content produced by members of the LAPC and invited authors followed each author's personal interests and areas of expertise although a general focus on climate change was a constant. Some respondents explained that writing for the website allowed them to reach a larger audience for their work, and also helped them in extending their social network:

> Well, to me it seems that the greatest benefit is to spread part of my research, and as an activist and researcher who analyzes such issues ... obviously one produces information, but that information is useless if people do not know about it. (SG) Some people have approached me because they did not know who was the person who worked on this issue, somehow (the website) facilitated this contact with me, or I have seen that people who read the articles have published them in other spaces. (AR)

Responses collected about utility, benefits, and overall objectives of this website from the internal questionnaires carried out in coordination between the CES team and the FFLA as the LAPC's executive secretariat, as well as from the in-depth interviews, were mostly positive.

Despite the overall positive image of the website, most interviewees also pointed out some limitations about the content being produced. This is related to the difficulties of balancing diverse information unique to each one of the countries represented by the LAPC and the website:

The website needs to maintain a good balance and equilibrium between the authors, the technical contents, themes and all, to reflect the political and ideological diversity that exists in Latin America. (RF)

So, at times we focus too much on the national agenda and we ignore other regional issues, therefore I think the challenge is to position an agenda and draw the attention of readers, I think that's the biggest challenge in these types of spaces. (SG)

Interviewees were also asked to discuss the utility of the website as a source of climate change information in the region. The responses pointed toward a variety of perspectives, ranging from the discussion of the website as a premier source of climate change policy information, to a secondary source after other official sources (e.g. UN reports) and information distributed via their networks (e.g. email, listservs). For others, the website was not a source of information because climate change is not their area of expertise.

The limitations discussed above resulted largely from limited contributions by many of the LAPC members. Except for a few notable exceptions, from the project's inception in December 2010, members of the LAPC and the Executive Secretariat did not contribute consistently to *Intercambio Climático*, particularly in submitting content such as articles and other materials. Therefore, in March 2011, the CES team began investing time in inviting guest authors to contribute to the website. Initially some of these invited authors were from Brown University working with the authors. The majority, however, were climate change specialists from Latin America who were aware of the work of the LAPC or part of its extended network such as the Latin American branch of the Climate Action Network and Latin American country negotiators to the UNFCCC.

As of September 2012, 34 invited authors had contributed to the website compared to 12 members of the LAPC. Table 1 shows that the contribution of articles from members of the LAPC (22%) has been considerably lower than the total contribution of the invited authors (78%). One benefit of this is to strengthen the goal of creating a constructive and progressive website for dialog between the LAPC, climate change policy–makers, and representatives from civil society organizations working on similar areas to that of the LAPC.

As fewer LAPC members participated than originally expected, members of the CES team continued to manage and edit the website past the original date in 2011 laid out in the agreement between the CES and the LAPC to hand over the administration of the website. Once the website shifted the focus onto invited

**Table 1.** Authorship of all articles published in *Intercambio Climático* from December 2010 until September 2012 ($N = 275$).

| Authorship | Articles |
| --- | --- |
| LAPC members | 22% ($n = 61$) |
| Guest authors (non-Brown University) | 17% ($n = 46$) |
| Guest authors (Brown University) | 61% ($n = 168$) |
| Total guest authors (Brown and non-Brown) | 78% ($n = 214$) |

authors, the pressure on the FFLA and the Platform to deliver on their commitments was further reduced. Respondents repeatedly said that the main reason preventing them from regularly contributing content was the lack of time to write:

> I do not want to spend another minute of every hour necessary and essential that I already spend in front of my computer. The computer absorbs a lot more time than I would want to, to also add new communication channels. (RF)
> What happens is that all these are resources of time, people, and cost that take to keep a website. That is the limitation that I see for now, and the same if we go to social networks, there have to be people dedicated to that. (SL)

*Challenges in the development of communication strategies*

The interviews and the internal assessment conducted by the CES team in conjunction with the FFLA as executive secretariat reveal important limitations and challenges to the website, both internal (e.g. financial and organizational) and external (e.g. interest of governments):

> I think the platform is currently at a critical point. (SL)
> So it's a matter of capabilities, it is a matter of financial resources, it is sometimes a matter of policy barriers in each country. (SG)

These online communication tools still require an important financial commitment by the organizations (Kurniawan & Rye, 2014), which in some cases, can be perceived as a lower priority. The financial aspect is considered a necessary but insufficient condition for success.

Various attempts were made to establish better lines of communication between the CES team and the LAPC and FFLA for working on the website, but due to repeated personnel changes in the LAPC, lack of time committed to the website activities by the FFLA, and lack of time by the LAPC members to contribute material, the project never really got past the initial stages.

Finally, a fundamental weakness from the outset was that the original conception of the project came from the CES team and not the LAPC. Although the partnership and initial planning and building of the site was conducted principally through direct meetings between the CES and the LAPC with a specific timetable for handing over the website to the LAPC included in the agreement, the LAPC never took real ownership of the website. This is despite a number of attempts by the CES team to train the FFLA team and transfer responsibilities. Personnel changes in the FFLA in particular undermined this process. Some interviewees, as shown below, recognized this issue:

> the appropriation of the space (website) has not occurred, and even though we set up goals in the [LAPC annual] assembly such as: "Each member write two articles per year," that did not quite happen. (RF)

**Discussion and Conclusion**

This case study focused on the development, accomplishments, and challenges of a unique collaboration among NGOs working on climate change issues in Latin

America with a US-based partner, and the use of online media to achieve its goals. The results of this study suggest that generally the participants in the LAPC perceive the website positively, but there are also problematic aspects of the collaboration that are not fully recognized by the members, which prevent a more functional and effective communication strategy.

The LAPC and the website have now completed the stage of development and are now entering a period of relative maturity. Interviewees recognize this as both a challenge and a substantial opportunity, especially in the next few years, considering that the COP20 in 2014 took place in Lima, Peru:

> The next challenge is that this [platform] will not fade away quickly, because the truth is that the regional networks and platforms, what experience shows is that you have a peak period of creativity, enthusiasm, and have a period of wear also. (PF)

The study revealed that the limitations discussed in previous studies also constrained the potential impact of the website. Financial and time constraints, lack of ownership and identification with a shared identity, were described during the interviews. Given the lack of involvement of the LAPC members, it is the authors' contention that clearer goals would probably not have changed very much and the idea to invite new authors might not have been realized, which as mentioned above has been one of the inadvertent success stories of this collaboration. However, the issues related to member participation could have been avoided with a feasibility study, which was not conducted to access whether the proposed website strategy would actually work in practice.

Returning to our initial question, can online forums support NGOs in addressing climate change? This case study left us with not only some positive experiences and directions for future improvements, but also some sobering realities. These realities include the need to effectively plan and manage expectations for these types of collaborations and the difficulty of sharing the responsibility for the maintenance of a website with fresh material across a network of 17 NGOs in 10 countries. In this respect, we have some final very concrete steps that might be taken to overcome some of the barriers *Intercambio Climático* has encountered, and which could be applicable to current and future initiatives like this one. Institutional limitations of the LAPC suggest that the lack of resources is a major limitation of the website. One alternative to improve the overall communication strategy of the LAPC is with a better-orchestrated media approach via the establishment of a dedicated team of communication professionals. More specifically, it is necessary to contract a full-time trained editor and/or website manager who can dedicate their time to the website. This could allow for the expansion and consolidation of *Intercambio Climático* and support the expansion of the LAPC's reach, as discussed by this interviewee:

> I think [the challenge] is to get media exposure at extra-continental scale. I truly believe that once the domestic, national communication have been served, and regional activities completed, the platform should seriously consider becoming a

serious source, responsible, longed, coveted, valued by CNN, BBC, EFE, Radio France International. (AM)

However taking the step of hiring such staff would also diminish the shared nature of the burden of collective organization, and may result in further distancing of member groups from the site's day-to-day success. In addition, the role of social media needs to be better integrated into the planning of the LAPC and its members. Social media tools such as Twitter and Facebook are part of the communication repertoire of the LAPC, but serve a limited and marginal function despite the *Intercambio Climático* website acting as the LAPC's de facto website. However, despite the overall positive viewpoint expressed by several interviewees, some important barriers and limitations were discussed:

> in the strict terms of work, (the use of social media) creates more work, but also has its advantages in that if one gets to really use it effectively and conscientiously, planned, it can have very good results. (EM)

Given the key role of the invited authors, the LAPC should focus sufficient time and effort to ensuring the ongoing participation of these individuals. Searching for new guest authors from the region and integrating them into the website can support the regular updating of the website with new content. It is then important to establish realistic goals for new content from LAPC members and guest authors, and to set targets for increasing the number of followers and fans on Twitter and Facebook, respectively.

Many of the impacts of the site may never be known, because its users remain hidden behind IP addresses. Without more interactive features, *Intercambio Climático* remains only a partial test of the ability of online forums to support civil society. The results of this study are aligned with those of previous studies that suggest that online media have not radically changed the communication efforts of NGOs (Lester & Hutchins, 2009; Pickerill, 2001). Additionally, in a context of lower Internet penetration, a large number of individuals are still marginalized from the conversation. Nevertheless, our results suggest that the availability of the information via the website was an important way for NGOs to engage in policy decisions. The results support findings emerging for research on NGO influence carried out by Betsill and Corell (2001). Future research will hopefully be able to learn from this project, and we hope that claims about the marvels of the Internet for organizing engagement will be balanced with real case studies such as this one.

## Notes

1. Figures from the end of 2013 indicated that Internet users in the region number approximately 296 million or 49.9% of the region's population. See http://www.internetworldstats.com/stats10.htm
2. See http://intercambioclimatico.com/en/ for the English language side of the site. There is different content on the Spanish and Portuguese sites at that address.
3. See http://intercambioclimatico.com/en/about-lapc/ for a list of participating organizations
4. http://www.intercambioclimatico.com/en/about-lapc/ (Accessed October 21, 2013).

5. *Intercambio Climático* "The Platform launches reports on climate change policies in 10 Latin American countries" Intercambio Climático website, November 14, 2012.
6. Alison Kirsch y Enrique Maurtua "CANLA y PCL exponen Políticas Públicas de Latinoamérica para Cambio Climático" 22 November, 2013.
7. Guy Edwards "Peru and Venezuela compete to host COP20 in 2014" Intercambio Climatico, 13 February, 2013 http://intercambioclimatico.com/en/2013/02/13/peru-and-venezuela-compete-to-host-cop20-in-2014/
8. Guy Edwards "Latin American Civil Society Organizations back Peru's bid to host COP20" Intercambio Climatico, June 5, 2013 http://intercambioclimatico.com/en/2013/06/05/latin-american-civil-society-organizations-back-perus-bid-to-host-cop20/ Accessed June 23, 2014.
9. Authors' personal communication with webmasters in September, 2012, of the following websites AlertNet's Climate Change, Climate and Development Knowledge Network, and Responding to Climate Change.

## References

Ackland, R., & O'Neil, M. (2011). Online collective identity: The case of the environmental movement. *Social Networks, 33*(3), 177–190. doi:10.1016/j.socnet.2011.03.001

Bach, J., & Stark, D. (2004). Link, search, interact: The co-evolution of NGOs and interactive technology. *Theory, Culture & Society, 21*(3), 101–117. doi:10.1177/0263276404043622

Bennett, W. L. (2012). The personalization of politics: Political identity, social media, and changing patterns of participation. *The ANNALS of the American Academy of Political and Social Science, 644*(1), 20–39. doi:10.1177/0002716212451428

Betsill, M. M., & Corell, E. (2001). NGO influence in international environmental negotiations: A framework for analysis. *Global Environmental Politics, 1*(4), 65–85.

Betzold, C. (2010). 'Borrowing' power to influence international negotiations: AOSIS in the climate change regime, 1990–1997. *Politics, 30*(3), 131–148.

Bortree, D. S., & Seltzer, T. (2009). Dialogic strategies and outcomes: An analysis of environmental advocacy groups' Facebook profiles. *Public Relations Review, 35*, 317–319.

Boykoff, M. T. (2009). We speak for the trees: Media reporting on the environment. *Annual Review of Environment and Resources, 34*, 431–457. doi:10.1146/annurev.environ.051308.084254

Brunsting, S., & Postmes, T. (2002). Social movement participation in the digital age: Predicting offline and online collective action. *Small Group Research, 33*, 525–554. doi:10.1177/104649602237169

Carvalho, A., & Peterson, T. R. (2012). *Climate change politics: Communication and public engagement.* Amherst, NY: Cambria Press.

Corell, E., & Betsill, M. M. (2001). A comparative look at NGO influence in international environmental negotiations: Desertification and climate change. *Global Environmental Politics, 1*(4), 86–107.

Cox, R. (2012). *Environmental communication and the public sphere.* Thousand Oaks, CA: Sage.

Creswell, J. W. (2012). *Qualitative inquiry and research design: Choosing among five approaches.* London: Sage.

Edwards, G. (2013). *The politics of climate change in Latin America: Leaders and laggards.* Retrieved from http://intercambioclimatico.com/en/2013/01/25/the-politics-of-climate-change-in-latin-america-leaders-and-laggards/

Edwards, G., & Roberts, J. T. (2015). *Leaders from a fragmented continent: The global politics of climate change in Latin America.* Cambridge, MA: MIT Press.

*El Cambio Climático No Tiene Fronteras.* (2008). Lima: Comunidad Andina.

*¿El Fin de las Cumbres Nevadas? Glaciares y Cambio Climático en la Comunidad Andina.* (2007). Lima: Comunidad Andina.

Endres, D., Sprain, L. M., & Peterson, T. R. (2009). *Social movement to address climate change: Local steps for global action.* Amherst, NY: Cambria Press.

Feldpausch-Parker, A. M., Parker, I. D., & Peterson, T. R. (2012). 350.org: A case study of an international web-based environmental campaign. In A. Carvalho & T. R. Peterson (Eds.), *Climate change politics: Communication and public engagement* (pp. 211–242). Amherst, NY: Cambria Press.

Gough, C., & Shackley, S. (2001). The respectable politics of climate change: The epistemic communities and NGOs. *International Affairs, 77,* 329–346.

Humphreys, D. (2004). Redefining the issues: NGO influence on international forest negotiations. *Global Environmental Politics, 4*(2), 51–74.

Intercambio Climático. (n.d.). About LAPC. Retrieved September 10, 2014, from http://inter-cambioclimatico.com/en/about-lapc/

Kent, M. L., Taylor, M., & White, W. J. (2003). The relationship between web site design and organizational responsiveness to stakeholders. *Public Relations Review, 29*(1), 63–77.

Kurniawan, N. I., & Rye, S. A. (2014). Online environmental activism and Internet use in the Indonesian environmental movement. *Information Development, 30,* 200–212. doi:10.1177/0266666913485260

Lester, L., & Hutchins, B. (2009). Power games: Environmental protest, news media and the internet. *Media, Culture & Society, 31,* 579–595. doi:10.1177/0163443709335201

Martins, A. (2011, November 16). América Latina, "la menos escéptica sobre el cambio climático". BBC Mundo. Retrieved from http://www.bbc.co.uk/mundo/noticias/2011/11/111115_clima_escepticismo_am.shtml

McCright, A. M., & Dunlap, R. E. (2003). Defeating Kyoto: The conservative movement's impact on U.S. climate change policy. *Social Problems, 50,* 348–373. doi:10.1525/sp.2003.50.3.348

Newell, P. (2008). Civil society, corporate accountability and the politics of climate change. *Global Environmental Politics, 8*(3), 122–153.

Pew Research Center. (2013). *Climate Change and Financial Instability Seen as Top Global Threats* [Survey Report]. Retrieved from http://www.pewglobal.org/2013/06/24/climate-change-and-financial-instability-seen-as-top-global-threats/

Pickerill, J. (2001). Environmental internet activism in Britain. *Peace Review, 13,* 365–370. doi:10.1080/13668800120079063

Postmes, T., & Brunsting, S. (2002). Collective action in the age of the internet: Mass communication and online mobilization. *Social Science Computer Review, 20,* 290–301. doi:10.1177/089443930202000306

Ryan, D. (2012). *Informe sobre el Estado y Calidad de las Políticas Públicas sobre Cambio Climático y Desarrollo en América Latina: Sector Agropecuario y Forestal.* Plataforma Climática Latinoamericana. Retrieved from http://intercambioclimatico.com/wp-content/uploads/Informe-regional-final-oct.pdf

Takahashi, B., & Meisner, M. (2013). Agenda setting and issue definition at the micro level: Giving climate change a voice in the Peruvian Congress. *Latin American Policy, 4,* 340–357.

Taylor, M., Kent, M. L., & White, W. J. (2001). How activist organizations are using the Internet to build relationships. *Public Relations Review, 27,* 263–284.

Wall, M. A. (2007). Social movements and email: Expressions of online identity in the globalization protests. *New Media & Society, 9,* 258–277. doi:10.1177/1461444807075007

Waters, R. D., Burnett, E., Lamm, A., & Lucas, J. (2009). Engaging stakeholders through social networking: How nonprofit organizations are using Facebook. *Public Relations Review, 35*(2), 102–106.

Xie, L. (2011). China's environmental activism in the age of globalization. *Asian Politics & Policy, 3,* 207–224. doi:10.1111/j.1943–0787.2011.01256.x

Yin, R. K. (2009). *Case study research: Design and methods* (vol. 5). Thousand Oaks, CA: Sage.

## Appendix: Position and institutional affiliation of interviewees during the time of interview

1. Representative in Patagonia Region, Fundación AVINA, Argentina (RF)
2. Director, EcoCom, Bolivia (AM)
3. Director ejecutivo, Sociedad Peruana de Derecho Ambiental, SPDA, Peru (PF)
4. Member of climate change and energy group, Centro Uruguayo de Tecnologías Apropiadas, CEUTA, Uruguay (SL)
5. International Policy Advisor, Fundación Biosfera y Climate Action Network Latin America (CAN-LA), Argentina (EM)
6. Legal advisor for climate change, Asociación Interamericana para la Defensa del Ambiente, AIDA, Bolivia (AR)
7. Legal specialist for the strategy of low emissions development in Mexico, Programa de Clima y Energía de WWF, Mexico (SG)

# Mobilizing Facebook Users against Facebook's Energy Policy: The Case of Greenpeace Unfriend Coal Campaign

Merav Katz-Kimchi & Idit Manosevitch

*This research analyzes Greenpeace International Unfriend Coal protest campaign against Facebook's energy policy (2010–2011), as a case study in the organization's approach to campaigning on climate change issues. In the context of Greenpeace protest history and social movement research, we focus on the online mobilization tactics used by Greenpeace, and how they differed from previous campaign practices. The findings are based on a content analysis of all statuses posted on the campaign Facebook page (N = 119), and six semi-structured interviews with key Greenpeace personnel. The analysis reveals that Greenpeace used Facebook extensively both for disseminating information and context about the campaign, and as a platform for mobilizing and facilitating broad public engagement. Greenpeace seized the affordances of the Facebook platform and introduced new means of online mobilization and engagement. These e-tactics proved effective for engaging diverse transnational supporters in the campaign. The implications for environmental non-governmental organizations and the broader environmental movement are discussed.*

## Introduction

Environmental non-governmental organizations (ENGOs) across the globe increasingly expand their use of digital technologies, viral marketing, and "prosuming" tactics, thereby facilitating dramatic changes in the nature of their protest campaigns (e.g., Bennett & Segerberg, 2012; Earl & Kimport, 2011). For example, ENGOs have been spreading YouTube videos to raise awareness of environmental causes (Stoddart & MacDonald, 2011), or to mobilize supporters to email US Senators to exert mass public pressure to help enact environmental policies (Hestres, 2014). ENGOs have also used cell phone text messages to coordinate demonstrations (Cullum, 2010) or to mobilize supporters to join on-the-ground protests (DeLuca, Sun, & Peeples, 2011). With the surge of social media, ENGOs have begun operating Facebook fan pages to provide contextually relevant information and links to further environmental resources (Stoddart & MacDonald, 2011; Netrebo, 2012).

Online digital tools lower the costs needed to mobilize supporters (Earl & Kimport, 2011), sustain networks of activists (Lim, 2012), and diversify the tactical repertoires of movements (Earl & Kimport, 2011; Garrett, 2006). However, the power of web 2.0 applications lies far beyond convenience, budget, and diversification. Interactive web 2.0 applications, specifically digital social networks, open up new opportunities for mobilizing broad publics to take an active role in environmental campaigns (della Porta & Mosca, 2006). In the context of journalism, Gillmor (2004) argues that the public has turned from a passive audience to an active participant in the conversation. Similarly, in the context of social movements, it seems that lay supporters have transformed from passive supporters to active participants in the campaign (Bennett & Segerberg, 2012; Earl & Kimport, 2011).

This study focuses on public mobilization in environmental campaigns in the context of ENGOs' global effort to promote "public engagement with climate change" (Whitmarsh, O'Neill, & Lorenzoni, 2013). We follow scholars who argue that the deployment of online digital tools has transformed the nature of social movements and collective action (Bennett & Segerberg, 2012; Earl & Kimport, 2011), and focus on the role of the lay public in campaigning—how it has changed, and the implications of these changes for environmental campaigning in general and climate campaigning in particular.

Our case study is the Greenpeace Unfriend[1] Coal campaign directed against Facebook's energy policy—one of the first protest campaigns in which Greenpeace extensively deployed social media as a major tactic to mobilize transnational supporters to engage in wide-ranging contentious protest activism. Greenpeace is an important ENGO for inquiry because it is considered the most prominent and media savvy environmental group (Carvalho, 2010; Dale, 1996), and a trailblazer in its innovative usage of digital media (Castells, 2009; Lester & Hutchins, 2009).

The goal of this study is twofold. First, our aim is to empirically investigate the affordances of social media to mobilize broad publics to engage actively in environmental campaigns. We focus on Facebook because it is the dominant social network today (Ray, 2013). To this end, we analyze the ways in which Greenpeace used its Unfriend Coal Facebook page during the year 2011. A qualitative content analysis was conducted of all the statuses posted by Greenpeace on the campaign Facebook page ($N = 119$). The analysis details the types of e-tactics Greenpeace used to mobilize the support of Facebook users, of whom one-third were also registered

Greenpeace supporters. We support our analysis by drawing on six semi-structured interviews with Greenpeace personnel. Second, at the theoretical level, this study aims to explore the significance of the empirical findings as regard ENGOs' campaigning for climate change, and specifically the role of the public in these endeavors.

The paper begins with theoretical background on environmental campaigning and mobilization, and the changes that the Internet has brought to this field. We then turn to a detailed description of our case study, first Greenpeace and then the evolvement of the Unfriend Coal campaign. Next, the analysis section presents the main e-tactics manifested in the statuses posted on the Unfriend Coal campaign page. We conclude by discussing the changes brought by social media to Greenpeace campaign tactics, and the broader implications of the interplay between the organization, the news media, and the public in promoting climate change causes.

## Theoretical Background

### *The environmental advocacy campaign*

An advocacy campaign is a strategic course of action that involves communication, and is undertaken for the purpose of achieving a specific goal during a pre-defined time frame (Cox, 2013, p. 213). Environmental advocacy campaigns are waged by non-institutional organizations, mostly ENGOs such as the Greenpeace campaign against Nestlé palm oil sourcing practices (Harrild, 2010). In some cases, these campaigns seek to change external conditions, for example stopping the construction of a hydroelectric dam on the Jordan River by the Society for the Protection of Nature in Israel (de-Shalit, 2001); in other cases, they aim for more systemic changes such as transforming corporate energy policies by the Sierra Club "Beyond Coal" campaign (Cox, 2010).

An advocacy campaign begins by defining an overarching *strategy*: "the crucial source of influence or leverage to persuade a primary decision maker to act on a campaign's objective" (Cox, 2013, p. 220). Once a strategy is defined, the organization plans the *campaign tactics,* i.e., the concrete actions designed to implement the broader strategy (Cox, 2013). To illustrate, in the Beyond Coal campaign of the Sierra Club, the strategy was to pressure legislators to prohibit the construction of new power plants, whereas a primary tactic was petition signing (Cox, 2010). With the advent of the Internet, the repertoire of campaign tactics expanded by introducing a wide array of *e-tactics*—collective actions that involve varying degrees of online components. These include online petitioning, email writing, online letter writing, and more (Earl & Kimport, 2011).

Environmental advocacy campaigns are inherently conflict situations that involve multiple "players" who differ in identity and interests, and consequently in the role they play in the campaign lifespan (della Porta & Rucht, 2002, p. 3). One key player in the campaign is the *primary audience*, the decision-makers authorized to act upon and implement the campaign objectives (Cox, 2013). The other key player is *the secondary audience* of the campaign, which consists of the various segments of the public, coalition partners, opinion leaders, and news media organizations that may hold the primary audience accountable for the campaign's objectives (Cox, 2013).

ENGOs employ a multi-faceted approach to campaigning, where several sets of tactics are designed, each of which is crafted to appeal to a different player in the campaign. Campaigns usually approach the primary audience directly by making an explicit formal request for a change in policy or routine. A direct appeal is no doubt an imperative act of communication that defines the complexity of the issue, and should ultimately lead to the crafting of new policies. But rational scientific persuasion of office holders has proven insufficient. Effective campaigning requires broad public support from constituencies that demand accountability from decision-makers, and in essence forces them to implement actual change (Cox, 2013).

The various segments of the secondary audience are therefore crucial players in environmental campaigns. ENGOs have made extensive, concerted efforts to garner support among various segments of the secondary audience. At the minimum, campaigners seek to mobilize the public for general agreement with the campaign objectives. This type of mobilization is known as consensus mobilization (Edwards & McCarthy, 2006) and can serve as indirect pressure on decision-makers to implement the requested changes. But more often, ENGOs seek to mobilize the secondary audience for *action mobilization* (Edwards & McCarthy, 2006), which may translate into explicit public outcries for the campaign objectives, and may lead constituencies to actively protest and demand accountability from decision-makers (Cox, 2013).

## Campaign mobilization

Political communication theorists identify three key components of mobilizing citizens for action: contextual information, empowerment, and mobilizing structures. Contextual information places the cause in a general context that provides a perspective for assessing the validity and importance of the claim (Iyengar, 1994). When appealing for environmental change, appropriate scientific information is necessary because most citizens do not have the knowledge needed to evaluate the urgency and necessity of change. Other important informational components are details of policies and stakeholders, the range of possible solutions, and the relative costs and benefits of each one (de-Shalit, 2001).

*A sense of empowerment* is a political consciousness that reflects people's belief that they can make a difference. Such a disposition helps drive constituents to try to change the political system (Gamson, 1992). Empowering rhetoric can be general and vague, such as "we can do it," or "yes we can." But rhetoric can also provide information that prods citizens to act (Lemert, 1977), for example, information about upcoming events such as massive virtual sit-ins where supporters block servers of relevant targets.

Finally, *mobilizing structures* are actual mechanisms that enable supporters to organize and engage in collective action. These include social structures and tactical repertoires. Social structures encompass formal configurations such as social movement organizations, and informal configurations such as activist networks.

Tactical repertoires refer to the forms of protest and collective action that activists are familiar with and able to utilize (Taylor & Van Dyke, 2009). Garrett (2006) suggests that supporters are more likely to mobilize around an issue if there is an existing organizational infrastructure and familiar forms of protest or contentious repertoires.

*Mainstream media and campaign mobilization*

Traditionally, mainstream news media served as the primary means for raising awareness and understanding of a cause, and mobilize the secondary audience for action. Extensive and sympathetic media coverage was considered a sine qua non for securing successful protest activity of social movements (Gamson & Wolfsfeld, 1993; Ryan, 1991), the environmental movement included (Hansen, 2010). The dependence of ENGOs on mainstream media largely shaped their daily operations (Hansen, 2010). The centrality of the news media organizations and their demand for photographic drama pushed radical environmental groups like Greenpeace to stage "image events" for mass media dissemination using still photography or later, video footage (DeLuca, 1999, 2011). The purpose was to invite the general public to bear witness on a mass scale by providing first-hand visual accounts of contentious activity, usually intense confrontations between professional Greenpeace activists and their opponents. ENGOs provided mainstream media with drama, conflict, action, and colorful photos (Gamson & Wolfsfeld, 1993). In turn, the mainstream media helped place environmental issues on the public agenda.

But the dependence on mainstream media we argue had deeper implications than resource allocation and work routines. It forced ENGOs to adopt a particular understanding of the nature of the campaign, and the interplay between the players involved. Lay citizens were generally attributed the role of passive supporters (Eden, 2004; Eyerman & Jamison, 1989) and fundraising.[2] This approach characterized the pragmatic business-oriented approach of prominent ENGOs like Greenpeace that focused on developing effective means for achieving outcomes in a timely manner (Castells, 1997, p. 118). In short, ENGOs could not rely on lay citizen activism to achieve their goals. Instead, they hired professional staff as the main activists (Eden, 2004).

*Digital platforms: new directions for environmental campaigns*

The rise of the Internet changed this state of affairs. As an alternative public sphere, the Internet enables ENGOs to maintain direct communication with the public, thereby liberating them, at least potentially, from complete dependence on the mass media as a go-between. ENGOs increasingly use web applications to promote visibility, disseminate information and context, raise public awareness and accrue legitimacy by using direct communication via websites, Facebook pages, Twitter accounts, YouTube videos, etc. (Stoddart & MacDonald, 2011).

The affordance of web communication, however, goes beyond the benefits of direct communication. Sharply reduced costs for organizing and participating in protest activities, along with ability to aggregate individual actions into broader collective actions, are two primary affordances associated with recent changes in environmental campaigning (Earl & Kimport, 2011). The Internet has broadened the repertoire of protest tactics that environmental campaigns can execute (Carroll & Hackett, 2006; Pickerill, 2003) by introducing *e-tactics* as innovative forms of collective action (Earl & Kimport, 2011). By overcoming the constraints of cost, time, and geographic location, the Internet has paved the way for ENGOs to use simple tactics to achieve *broad* public participation.

Early usage of the Internet for environmental campaigning was based on email and website platforms. ENGOs created unofficial spoof websites of their opponents, or provided platforms for online petitioning and letter writing (Pickerill, 2003). With the escalating popularity of social network platforms, social movements have recently begun to integrate social media in their campaigning (Harlow, 2012; Stoddart & MacDonald, 2011).

However, the integration of digital social networks, in particular, has not only altered the means of social campaigning and the scope of public participation. Rather web-based digital media have transformed the nature of collective action (Bennett & Segerberg, 2012; Earl & Kimport, 2011). Bennett and Segerberg (2012) argue that this change stems from a new logic of action that underpins much of the large-scale social protests in recent years. They call this the *logic of connective action*, which they differentiate from the traditional logic of *collective action*. The latter is based on the creation of a collective identity, and therefore a priori limits the scope of participation, and hence requires high levels of organizational resources. In contrast, the logic of connective action is based on the recognition of digital media as organizing agents.

Connective action is self-motivated by the desire to share "already internalized or personalized ideas, plans, images, and resources with networks of others" (Bennett & Segerberg, 2012, 753). Thus, contributing to a common cause does not express belongingness to the collective identity of a particular social movement. Rather such participation becomes an act of "personal expression and recognition or self validation," and it is achieved by "sharing ideas and actions in trusted relationships" (pp. 752–753). As such, the logic of connective action does not entail collective identity; rather it facilitates accessible participation opportunities of diverse audiences, with fairly minimal organizational resources. Our case study focuses on an "organizationally enabled" campaign in which formal organizations like Greenpeace "soft-pedal demands for formal membership, as well as collective identity and action framing" (Bennett & Segerberg, 2013, p. 48).

The dominance of Facebook in the social networking arena today, totaling 1.15 billion users (Facebook, 2013; Ray, 2013), makes it a particularly important platform for campaigning. Scholars have begun to examine the use of Facebook by ENGOs. For example, Stoddart and MacDonald (2011) demonstrated how Facebook pages can encourage participation in ENGOs' protest activities that were modeled after traditional forms of activism (e.g., information on ways to get campaign bumper stickers), but did not generate new forms of protest. More broadly, Harlow (2012) examined how users' comments were framed to mobilize and advance an online social justice movement that activated an offline movement.

Bennett and Segerberg (2012) provided a pioneering macro-level analysis of the organizational aspects of connective actions by focusing on diverse social justice and environmental movements. Their analysis presents an intriguing new conceptualization of social protest, and helps explain the unprecedented effectiveness of digital networks in campaigning. However, further work is needed to better understand the micro-level dynamics of connective action, and the specific ways in which it manifests itself in practice. Clearly, the integration of the Internet in environmental campaigns, and in particular the use of interactive digitally connected social networks, has paved the way for easy e-participation that potentially changes the role of supporters of a campaign.

In what follows we examine the employment of social media in one specific climate change campaign. Our driving research question is: in what ways did Greenpeace campaign organizers use the Facebook platform to mobilize broad support and participation in their Unfriend Coal campaign? Specifically, we examine the type of information disseminated by Greenpeace on the page, and the types of e-tactics employed. Our goal is to reveal novel means of utilizing social media for campaign purposes, and to consider what these tools imply with regard to changes in Greenpeace's approach to campaigning and specifically to the role of the public in their campaigns, and the broader implications for climate campaigning.

## Greenpeace and Climate Change Communication

Greenpeace, founded in 1971, has been campaigning about anthropogenic climate change since the late 1980s (Boykoff, 2011; Doyle, 2007, 2011; Hansen, 1993; Lipschutz & McKendry, 2011; Yearley, 2008), by communicating key scientific findings to "skeptical governments and a disinterested public" (Doyle, 2007, p. 129), and later images of nature at risk (Doyle, 2007, 2011). However, this campaign tactic did not prove successful because it did not attract the attention of mainstream news media that are primarily interested in event-based news (Doyle, 2007).

Since the early 2000s, Greenpeace has expanded its campaigning strategies and tactics regarding climate change. Against the backdrop of accelerated globalization (Bauman, 1998), the failure of international attempts to regulate transnational corporations (TNCs) and the withdrawal of the nation-state from enforcing certain regulatory functions with regard to TNCs and climate change in particular (Newell, 2000a), Greenpeace began targeting TNCs themselves (Newell, 2000b). For example, since the early 2000s, Greenpeace has targeted oil companies such as Esso/ExxonMobil to encourage them to change their stance on climate change (Doyle 2011; Eden, 2004).

This change in campaign targets has been accompanied by e-tactics, thus demonstrating the effective harnessing of new media technologies for mobilization and protest campaigns. For example, in 2000 Greenpeace initiated Cokespotlight campaign against Coke Cola, a key sponsor of the Sydney Olympics, by creating a website designed to protest against the company's use of hydrofluorocarbons, highly potent greenhouse gases, as refrigerants in its vending machines (Warkentin, 2001). The website was used to mobilize Internet users across the globe to actively protest by downloading campaign stickers, posters, and postcards. Further, users were encouraged to lobby Coca-Cola directly, by providing email templates to write to the company's CEO. The campaign successfully ended in 2004 when Coca-Cola switched to other refrigerants.[3] A later example was Green-my-Apple campaign launched in 2006 against Apple's use of toxic chemicals in its products. While not related to climate change, this campaign illustrates the way in which Greenpeace expanded the possibilities for online activism. Through the designated website, users could email Apple's CEO, easily recommend the website to social networking services like del.icio.us, blog about the campaign, and create poster and t-shirt designs. This website was popular, and the campaign proved successful when Apple declared a phase-out of the toxic chemicals in its products by 2008.[4]

Cokespotlight and Green-my-Apple campaigns marked the beginning of a change in Greenpeace's approach to campaigning; a change in the tools used, but more significantly, a change in the notion of the secondary audience and the ways in which it participates in a campaign. In both cases, Greenpeace developed online tools addressing a broad community of Internet users rather than its paying members alone. Further, in both cases, the organization used interactive features within its website to engage the broad transnational public in campaign activities. Differently put, the breadth of the secondary audience seems to have changed as well as its role in the campaign.

However, while website platforms are limited as regards their e-campaigning capabilities, social media platforms introduce new possibilities for mobilizing broad international publics. An important example is the Unfriend Coal campaign, which is the first Greenpeace campaign implemented via a Facebook fan page.

## The Case Study: Unfriend Coal Campaign

Greenpeace Unfriend Coal campaign was directed against the social media giant and took place over a 22-month period (February 2010 to December 2011). This was the first time that Greenpeace International implemented a comprehensive online contentious protest campaign via a designated Facebook page.[5] Greenpeace demanded that Facebook stop using coal as a main energy source in operating its data centers and switch to renewable energy sources such as wind or sun. The campaign successfully ended when Facebook announced its plan to use renewables in its data centers worldwide and collaborate with Greenpeace on implementing this plan.

The campaign was a culmination of two separate processes within Greenpeace. One was detailed analyses of the $CO_2$ emissions of cloud computing, which is part of the broader Greenpeace effort to tackle climate change (Rabinowitz, interview, September 2012). Based on a 2008 industry study, Greenpeace published a report on the CO2 emissions of cloud computing in March 2010 showing that cloud computing from the IT sector, including Facebook, contributes to a much larger carbon footprint than previously estimated. This report provided the justification and the incentive for targeting Facebook on the issue. The other process was a "digital comprehensive effort" within Greenpeace that started in 2005 concerning strategic possibilities and initiatives related to the digital world, recently including social media and smart phones (Weizmann, interview, March 2013).

The campaign officially started on February 2010 after revealing that Facebook planned to use coal as the main power source for its new data center in Oregon. The main strategy was to put pressure on Facebook's management because they were the primary audience authorized to implement changes in the company's energy policy. One important behind-the-scene tactic was direct negotiations and contacts with Facebook headquarters in Silicon Valley led by the Greenpeace San Francisco office (Harrell, interview, February 2013).

Two main social media platforms were used in addition to postings on the Greenpeace International website: YouTube and Facebook. YouTube helped the viral spread of short video clips. For instance, on 13 September, 2010, two weeks before the release of the film *The Social Network*, Greenpeace released its own version of Facebook's history in a sarcastic cartoon campaign video asking Mark Zuckerberg to unfriend coal.[6] The clip went viral, with Greenpeace offices around the world adding subtitles in different languages. The use of Facebook was initially decentralized. When the campaign was launched in February 2010, independent Facebook groups were created spontaneously by French, Dutch, and Spanish supporters, as well as an English-speaking group created by Greenpeace International (Harrell, interview, February 2013). Greenpeace estimates that by July 2010, when Facebook celebrated five million users, about 500,000 had joined Facebook groups associated with the Unfriend Coal campaign. These groups were active, creating momentum and facilitating ongoing e-tactics. But the decentralized structure of the campaign limited Greenpeace's ability to coordinate all supporters together for effective contentious activity.

On 21 January 2011, Greenpeace International launched the Unfriend Coal Facebook campaign page, thereby eliminating the weaknesses of the group pages and the decentralized organizational structure of the secondary audience.

## Research Design

Our research combined a content analysis of all Facebook statuses posted by Greenpeace administrators on their Facebook Unfriend Coal fan page during the year of 2011 ($N = 119$), and semi-structured interviews with key Greenpeace personnel ($N = 6$).[7] The content analysis was a three-step process. First, we conducted a quantitative content analysis to identify the extent to which Greenpeace leadership employed Facebook to communicate each of the two major components of campaign communication: contextual information and e-tactics. Each status was coded for three variables of interest. Statuses were coded on a dichotomous scale where "1" indicates that the status includes the measured variable, and "0" indicated that it did not include it.[8]

The first variable, *contextual information*, based on Iyengar's definition for thematic frames (1994), was coded as any information that places the cause in a general context. Since the cause is Facebook's energy policy, the general context consisted of any information about the energy policies and energy consumption of Facebook or the broader high-tech industry, their environmental effects, possible solutions, and their feasibility.

The second variable, *campaign tactic*, based on Cox's (2013) definition, was defined as any action designed to implement the main strategy of putting pressure on Facebook's management to stop using coal as a main energy source in operating its data centers and switch to renewable energy sources.

The third variable, *empowerment*, was defined as content suggesting that supporters can actively contribute to the campaign effort (based on Gamson, 1992, p. 7). Specifically, we coded all the content that reported about supporters' active involvement in the campaign.

Since this is an exploratory study, seeking to identify new types of e-tactics employed in the campaign, we supplemented our quantitative analysis with qualitative analyses of the statuses. After the initial coding of the statuses, we conducted a second analysis, looking for major themes within each type of content. Finally, we examined the statuses a third time, to identify any additional themes in the data that may shed light on novel ways by which Greenpeace utilized Facebook in this campaign. In the following we provide a detailed account of our findings. Note that the categories are not mutually exclusive, as each status may consist of several types of content.

## Analysis of Facebook Status Posts

*Descriptives*

Greenpeace administrators on the Unfriend Coal Facebook fan page posted a total of 119 statuses during 2011. Most statuses were posted during the initial months, in February ($n = 25$), March ($n = 20$), and April ($n = 34$). In subsequent months, the numbers ranged from 0 in July to 10 statuses in May.

Due to the nature of social media, we do not have an accurate account of the exposure to the page content, but several indicators provide a sense of the page impact. First, Greenpeace reports that a million unique users were involved in the online campaign, via Facebook as well as other online platforms during its 22-month duration (Harrell, interview, February 2013). Looking at tallies provided on Facebook, massive activity is apparent. Specifically, the Unfriend Coal Facebook page had a total of 184,622 likes (3 November 2013), which reflect the core number of users who choose to receive ongoing campaign updates on their feed. All statuses posted on the page received likes or comments, with the number of "likes" ranging from 10 to 2180 ($M = 370$), and the number of comments ranging from 1 to 280 ($M = 38$). One outlier status had 13,244 likes and 80,027 comments, when Greenpeace urged users to post comments to help attain a Guinness World Record of comments.

## An international endeavor

Greenpeace posted statuses in different languages including English, Turkish, French, Hebrew, Spanish, Swedish, Indonesian, and Italian, thereby directly addressing a diverse transnational public, and implicitly transmitting the message that this is an international endeavor (Harrell, interview, February 2013; Weizmann, interview, March 2013). Altogether, 103 posts were in English and 16 were written in other languages, some of which were duplicates of statuses reporting on key campaign events. This approach of mobilizing international audiences aligns with Greenpeace's strategy of targeting policy makers at TNCs (Newell, 2000b).

## Contextual information

*Facebook context.* Thirty-eight percent of the campaign statuses ($n = 45$) incorporated contextual information about Facebook. Two major themes emerge in the content of these postings. The dominant theme ($n = 35$) was information relating to Facebook's energy policies. A less dominant theme ($n = 11$) involved the company's business information, such as Facebook's market value (27 January 2011), and statistics related to the number of active users or shared items (5 February 2011).

*Broad context.* Forty percent of the statuses ($n = 48$) provided information pertaining to the broad context. The major theme ($n = 37$) was information pertaining to *energy policies and consumption of the IT sector at large*. For example, a report indicating that Microsoft "gets 25% of its annual energy from renewable sources, travels 30% less and boasts more energy-efficient buildings" (25 February 2011); or that Google has invested twice as much as Facebook in wind projects, and implemented a major solar project (16 April 2011). Greenpeace campaigner Casey Harrell was cited saying that electricity demand from the IT sector is the fastest growing segment in the USA (20 April 2011).

Less dominant contextual themes were *environmental, health, and economic implications of coal usage as energy source* ($n = 8$), and context about *Greenpeace's broad environmental activism* ($n = 4$), such as "at Greenpeace, we've been campaigning against nuclear for forty years" (March 17). These findings align with Schäfer (2012) who argues that ENGOs are champions in providing online information on climate change.

## E-tactics

Facebook's key design features, i.e., the ability to "like," "share," and "comment," are mobilizing structures embedded in the platform. These features provide users quick, easy, and familiar means to actively support the campaign, thereby adhering to theoretical conceptualizations of mobilizing structures (e.g., Lemert, 1977; McCarthy, 1996). Further, the Facebook culture and norms of use, namely, liking and sharing as an expected behavior, may serve as implicit mechanisms of encouraging users to go beyond lurking, and actively engage in the campaign.

Therefore, Greenpeace's choice to employ Facebook for the campaign is *in itself* an e-tactic. Not only does it provide a means for active online engagement, it also serves as an *implicit* means of empowerment and a call for action. And yet, this e-tactic may be reinforced by *explicit* content posted in statuses. Indeed, our content analysis of the statuses reveals that Greenpeace administrators supplemented these e-tactics with explicit content that they posted in statuses throughout the campaign. This content served as a call for action *as well as* a means of empowerment.

*Explicit call for action.* Nearly half of the statuses (*n* = 58, 49%) included an explicit call for action. The two most common types of a call for action were: a call to support the campaign, using the phrase "like our campaign page," (*n* = 15); and a call to share the campaign content, such as "pass it on" and "share" (*n* = 17). Less dominant were calls to *voice an opinion* (*n* = 8), for example, calling users to ask President Obama or Oprah Winfrey a campaign-related question in their live Facebook Q & A session during their visit to the company's headquarters (21 April 2011 and 8 September 2011, respectively); or, asking supporters to share their thoughts about Facebook's new energy initiatives, such as the partnership with Opower & NRDC (19 October 2011).

*Facebook events* were those occasions where Greenpeace initiated a campaign event. We call it a Facebook event since it was based on the affordance of the Facebook platform and was implemented on the campaign page. Several such events occurred during the campaign. Most visible was setting a Guinness World Record of Facebook comments (8 April 2011); and a photo competition where users were encouraged to demonstrate their support of the campaign by posting a self-portrait accompanied with the campaign graphic (11 February 2011). These Facebook events served primarily as a means of keeping supporters interested and engaged (Weizmann, interview, March 2013). Unlike former Greenpeace image events that involved few professional activists during a confrontational situation on challenging locations, Facebook events were conducted on a massive scale with lay people taking no actual risk.

Notably, this usage of social media and the corresponding notion of engagement differ from the approach employed by social movements in recent political campaigns, specifically the Arab Spring and the Occupy movement (e.g., Lim, 2012). In those cases, social media was employed to organize groups on the ground to take over urban territories, thereby linking online discourse and offline activism as two necessary components for achieving political goals. Interestingly, none of the statuses posted during the campaign called Greenpeace supporters to get out of the comfort zone of their social media disposition. E-tactics were web-based only, without any explicit expectation for offline activism. While Greenpeace leadership maintained ongoing campaign activity, with mobilizing opportunities throughout the year, all were confined to Facebook, usually requiring minimal contribution, such as showing support by clicking the "like" button, or passing campaign material on to their Facebook friends. These findings align with Earl and Kimport's (2011) findings that e-tactics are often easy-to-use engagement tools that demand minimal effort and are therefore effective in response. They also suggest a broad definition of engagement, which does not necessitate offline action, and considers online engagement as equally valuable. Indeed, online engagement proved effective. The cumulative commitment of the mass community of supporters, over the 22-month duration of the campaign, enabled ongoing pressure on Facebook.

*Empowerment*

The campaign was a long and dynamic strategic course of action. In order to maintain supporters interested and actively involved, Greenpeace campaigners continually reported about the campaign progress.

*Activism taken by supporters.* Nearly a quarter of the statuses ($n$ = 26, 22%) reported about campaign actions taken by users, such as the results of a campaign photo competition (11 February 2011), and a summary of the many activities taken by supporters, for example, "People have been joining the campaign in droves … in our photo competition … others donned our Facebook t-shirts and made videos. Plenty … passed along our new campaign video to your friends" (22 April 2011). Several statuses ($n$ = 9) provided reference to reports published in the media about the campaign progress, which may function as validation of the campaign's actions (Gamson & Wolfsfeld, 1993).

*Positive reinforcement approach to campaigning*

Using the Facebook platform for campaigning against Facebook may appear an ironic expression of protest, and an abuse of the company's product against it. But a close look at the Unfriend Coal page reveals the contrary. Throughout the campaign, Greenpeace statuses conveyed a clear message of unconditional support and even love of Facebook. The campaign was not undermining Facebook, nor was it overlooking its positive contribution to the world community.

Our study reveals that Greenpeace campaigners intentionally approached Facebook in a positive manner (Harrell, interview, February 2013), and this attitude was expressed in statuses throughout the campaign. Twenty-four percent of the statuses ($n$ = 29) manifested positive sentiments toward Facebook. For example, in a YouTube clip starring actors Ed Begley and Rachelle Carson, Carson said "we love Facebook," and Begley praised the company for being "fresh" and "innovative" (23 February 2011). In another status, Greenpeace congratulated Facebook—"We want to congratulate them on their progress"—when the company announced that its first data center outside the US would be built in Luleå, Sweden near the Arctic circle in order to use free cooling from the frigid local climate (27 October 2011). Three statuses positing Facebook with the challenge to "like renewable" also used positive rhetoric (e.g., 22 April 2011).

This campaign tactic seems to adhere to scholarship that underscores the qualities of positive reinforcement for promoting desired behavior. Adlerian psychology (named after Alfred Adler, 1870–1937) introduced the notion of *positive discipline* to describe an approach to child rearing in which care giving adults encourage and harbor feelings of adequacy, potency, and respect so that children are more responsive and cooperative (Kottman, 2001). Management studies (e.g. Daniels, 1999; Wexley & Nemeroff, 1975) use the term *positive reinforcement* to describe "any consequence that follows a behavior and increases its frequency in the future" (Daniels, 1999, p. 54). Positive reinforcement occurs when a behavior produces a favorable change in the environment for the performer and includes symbolic and tangible reinforcement such as praise, public acknowledgment, a plaque, or a trophy (Daniels, 1999).

Notably, none of the statuses call campaign supporters to boycott Facebook, or take any other type of action against the company. This is different from past campaigns that undertook a dissent and opposition approach toward the primary target (Castells, 1997; DeLuca, 1999).

## Conclusion

Since the birth of the modern environmental movement during the 1960s, lay citizens played an important, albeit mostly limited role in environmental campaigns, organized by ENGOs. Campaign organizers have traditionally relied on the news media for raising their causes to the public agenda, and mobilizing wide public consensus. While active citizen engagement was welcomed, many ENGOs, Greenpeace included, relied primarily on professional activists to execute their protest campaigns. The role of supporters was usually limited to passive public opinion and fundraising.

The penetration of the Internet has transformed this state of affairs by allowing direct communication with the public, thereby liberating ENGOs from complete dependence on mainstream media; and by providing a platform for mobilizing public support and active engagement in campaigns. Early usage of the Internet in environmental campaigns was limited to the technical possibilities afforded by website and email platforms. The rise of social media has opened up new opportunities for actively engaging broad publics in environmental campaigns.

Our analysis of the Unfriend Coal campaign serves as an important case study for successful employment of opportunities afforded by social media. Indeed, Greenpeace elaborated and perfected the employment of e-tactics, and mobilized an unprecedented number of global supporters in an environmental campaign. Three major findings emerge from our analysis.

First, Greenpeace used the campaign page for the dual purpose of continuously disseminating information and context about the campaign and its cause, as well as a mobilizing tool and a platform for broad public engagement. In essence, the campaign page served as a hub for all communication needs of Greenpeace and its supporters. But the employment of Facebook did not replace mainstream media or deter from the importance that campaign organizers attributed to it. Alongside the massive online activity via Facebook and other platforms Greenpeace campaigners continued to view mainstream media as an important agenda setting player, and a key tool for exerting pressure on the primary audience—Facebook management (Harrel, interview, February 2013). Accordingly, campaigners provided mainstream news media with frequent updates about the campaign.

It thus appears that Facebook has added a new dimension to Greenpeace campaigns, thereby serving as an additional medium that more likely complemented news media, rather than replaced it. The affordances of social media allowed Greenpeace to maintain ongoing communication with a broad international audience of supporters, and mobilize this audience for direct pressure on Facebook management. At the same time, the affordance of news media, as a non-affiliated bystander, helped amplify the cause reaching an even broader audience, including non-users, non-supporters, elites, and decision-makers. While content analysis of the press coverage is beyond the scope of this research, a search in Lexis-Nexis under the category of All News (English) reveals 142 news stories published in 2011, with 16 stories published in major newspapers worldwide. Much of this coverage was favorable to the Greenpeace cause. For example, in April 2011, upon the release of Greenpeace report *How-Dirty-Is-Your-Data*, the *Guardian* reported at length about the cause directly quoting the report (Carus, 2011). Other news stories reported about campaign e-tactics and the number of supporters (Parry, 2011; Roslin, 2011), and contextualized the campaign by referring to previous victories of Greenpeace (Pattenden, 2011) and other ENGOs (Kaufman, 2011).

Second, the Unfriend Coal campaign introduced a new line of e-tactics that were based on the affordances of the Facebook platform. The embedded options to like, share, and comment, were reinforced by rhetoric that called users to seize those options and engage, albeit from the comfort zone of their diverse locations across the world. Particularly interesting are the Facebook events that were implemented throughout the campaign. These e-tactics were created and designed based on the affordance of the platform, thus going beyond embedded means of engagement to novel usage of the medium. The significance of these findings is not only in the novelty of the tools but also in the role they took in the campaign. While earlier protest campaigns implemented by ENGOs, employed new media tools mainly for the purpose of attracting news media to the novelty of the tools (Lester & Hutchins, 2009; Pickerill, 2003), in the Unfriend Coal campaign, Greenpeace used the capabilities of social media primarily *for action mobilization and organizing* the broad public. News coverage that reflected on the novelty of tools was perceived as an added benefit of using social media rather than the goal of it.

Taken together, the use of Facebook not only transmitted a message that the public is an active participant, but actually enabled and encouraged such participation via empowering rhetoric and mobilizing technical features. The role of the public has changed from passive supporters to active participants. But here again, the employment of novel e-tactics and the broadening role of supporters, seemed to have complemented previous routines, rather than replaced them. Greenpeace did not forgo control of the campaign but "enabled" it (Bennett & Segerberg, 2013). The fan page was administrated by Greenpeace staff who carefully selected the content for the status posts, and actively monitored users' contributions. Further, the Facebook page was only one component of the broad campaign endeavor that included ongoing behind-the-scenes negotiations with Facebook management, as well as traditional off-line protest activities, such as flying the campaign airship above Facebook headquarters in Silicon Valley.

These findings seem to align with the logic of connective action recently introduced by Bennett and Segerberg (2012, 2013). Instead of projecting strong agendas and political brands, and allocating resources to building collective identities, Greenpeace used social media tools to enable loose public networks to form around personalized or individual action themes. By encouraging transnational users to seize social media opportunities for effective online engagement, Greenpeace seemed to have acted by the logic of connective action. This approach proved successful. Alongside traditional press coverage of the campaign in major world publications, the direct pressure of one million transnational unique users produced cumulative effect and eventually forced Facebook to collaborate with Greenpeace in creating new energy policies.

Study limitations should be noted. The content analysis of statuses is revealing as regards to the ways by which Greenpeace administrators approached the campaign. Complementing this data with interviews of key Greenpeace staff helps validate our findings. But further research is needed to complement this analysis, with an analysis of user content during the campaign. Another limitation stems from the nature of a case study. The success of the Unfriend Coal campaign helps illuminate the possibilities afforded by social media for ENGOs. But for generalization purposes, research needs to examine how Facebook is employed in other campaigns, to gain a broader perspective on the topics discussed above.

Looking ahead, this research may be useful to scholars and practitioners seeking to engage broad publics in environmental campaigns. A growing body of literature on climate change politics focuses on facilitating public engagement with climate issues (Nerlich, Koteyko, & Brown, 2010; Whitmarsh, O'Neill, & Lorenzoni, 2011, 2013). On the individual level, scholars identified three dimensions of engagement with these issues: cognitive (knowledge and understanding), affective (interest and concern), and behavioral (personal actions to minimize one's ecological footprint) (Lorenzoni, Nicholson-Cole, & Whitmarsh, 2007). Connective action campaigns of the type analyzed in this research induce cognitive and affective engagement but they do not demand behavioral engagement or personal change to decrease one's ecological footprint. In this sense, they are limited. But this limitation is also their strength. In an era of globalization, where commerce and production are often centralized by TNCs and businesses, environmental issues are often global. In this context, as explained in the theoretical section, Greenpeace aspires to impart a wide structural change and to promote increasing use of renewable energy resources by industry across the globe. Accordingly, Greenpeace targets TNCs, not individuals and their daily routines. Thus, connective action campaigns of the type analyzed here, may not achieve comprehensive individual engagement with climate issues. But they are particularly successful for bringing in a broad public support, for a single directive outcome. Social media platforms support the logic of connective actions and adhere to the characteristics of targeting a secondary audience.

## Acknowledgments

Both authors equally contributed to the article. We would like to thank Ori Tenenboim for his help in conducting the content analysis.

## Notes

1. In this article, we refer to Greenpeace International in Amsterdam unless otherwise stated.
2. An exception to this are a few offices of Greenpeace, such as Greenpeace UK, where supporters actively engage in promoting Greenpeace campaigns directed toward a local audience. Note that these groups are generally rare; they operate under the supervision of the local Greenpeace office, and do not have the autonomy to structure their campaign messages.
3. See   http://www.greenpeace.org/international/en/news/features/ask-and-ye-shall-receive-comp/ Retrieved September 2013.
4. See: http://members.greenpeace.org/sites/greenmyapple/, Retrieved September 2013. For additional information, see http://www.greenpeace.org/international/greeningofapple/ Retrieved September 2013.
5. The campaign page is available at: http://www.facebook.com/unfriendcoal?fref=ts. In early 2010, while campaigning against Nestlé, Greenpeace posted a provocative YouTube video that was spread virally on the Internet. The successful video helped Greenpeace assess the potential social media (Cohen, interview, April 2012).
6. The video is available at: http://www.youtube.com/watch?v=9q3pZVDHUJU
7. Interviewees were: Casey Harrell, campaign coordinator for the IT sector, Greenpeace International; Itamar Weizmann, former new media coordinator, Greenpeace Israel; Imrie Cohen, former energy campaigner, Greenpeace Israel; Danny Rabinowitz, former board member, Greenpeace UK; Tracy Frauzel, communication strategist, Greenpeace UK.
8. One coder coded the sample. A second coder coded 20% of the sample ($n = 26$) in order to determine inter-coder reliability levels for the measures. Agreement rates after correcting for chance using Scott's (1955) pi, were above 85%.

## References

Bauman, Z. (1998). *Globalization: The human consequences*. Cambridge: Polity.
Bennett, W. L., & Segerberg, A. (2012). The logic of connective action: Digital media and the personalization of contentious politics. *Information, Communication & Society, 15*, 739–768. doi:10.1080/1369118X.2012.670661

Bennett, W. L., & Segerberg, A. (2013). *The logic of connective action: Digital media and the personalization of contentious politics.* Cambridge: Cambridge University Press. doi:10.1017/CBO9781139198752

Boykoff, M. T. (2011). *Who speaks for the climate? Making sense of media reporting on climate change.* Cambridge: Cambridge University Press. doi:10.1017/CBO9780511978586

Carroll, W. K., & Hackett, R. A. (2006). Democratic media activism through the lens of social movement theory. *Media, Culture & Society, 28*(1), 83–104. doi:10.1177/0163443706059289

Carus, F. (2011, April 22). Apple's dirty data centers send it to bottom of green league: Yahoo first in Greenpeace's table of technology firms companies urged to cut reliance on coal power. *The Guardian.* Retrieved from http://www.theguardian.com/environment/2011/apr/21/apple-least-green-tech-company

Carvalho, A. (2010). Media(ted) discourses and climate change: A focus on political subjectivity and (dis)engagement. *Wiley Interdisciplinary Reviews: Climate Change, 1*(2), 172–179.

Castells, M. (1997). *The power of identity.* Oxford: Blackwell.

Castells, M. (2009). *Communication power.* Oxford: Oxford University Press.

Cox, R. J. (2010). Beyond frames: Recovering the strategic in climate communication. *Environmental Communication, 4*(1), 122–133.

Cox, R. J. (2013). *Environmental communication and the public sphere* (3rd ed.). Los Angeles, CA: Sage.

Cullum, B. (2010). Devices: The power of mobile phones. In M. Joyce (Ed.), *Digital activism decoded: The new mechanics of change* (pp. 47–70). New York: IDebate.

Dale, S. (1996). *Mcluhan's children: The Greenpeace message and the media.* Toronto: Between the Lines.

Daniels, A. C. (1999). *Bringing out the best in people.* New York: McGraw Hill, Inc.

DeLuca, K. (1999). *Image politics: The new rhetoric of environmental activism.* New York: Guilford Press.

DeLuca, K. (2011). Environmental movement media. In J. D. H. Downing (Ed.), *Encyclopedia of social movement media* (pp. 172–178). Los Angeles, CA: Sage.

DeLuca, K., Sun, Y., & Peeples, J. (2011). Wild public screens and image events from Seattle to China. In S. Cottle & L. Lester (Eds.), *Transnational protests and the media* (pp. 143–158). New York: Peter Lang.

de-Shalit, A. (2001). Ten Commandments of how to fail in an environmental campaign. *Environmental Politics, 10*(1), 111–137. doi:10.1080/714000516

della Porta, D., & Mosca, L. (2006). Global-net for global movements? A network of networks for a movement of movements. *Journal of Public Policy, 25*(1), 165–190. doi:10.1017/S0143814X05000255

della Porta, D., & Rucht, D. (2002). The dynamics of environmental campaigns. *Mobilization: An International Quarterly, 7*(1), 1–14.

Doyle, J. (2007). Picturing the clima(c)tic: Greenpeace and the representational politics of climate change communication. *Science as Culture, 16*(2), 129–150. doi:10.1080/09505430701368938

Doyle, J. (2011). *Mediating climate change.* Surrey: Ashgate.

Earl, J., & Kimport, K. (2011). *Digitally enabled social change: Activism in the internet age.* Cambridge, MA: The MIT Press. doi:10.7551/mitpress/9780262015103.001.0001

Eden, S. (2004). Greenpeace. *New Political Economy, 9*, 595–610. doi:10.1080/1356346042000311191

Edwards, B., & McCarthy, J. D. (2006). Resources and social movement mobilization. In D. Snow, S. A. Soule, & H. Kriesi (Eds.), *The Blackwell companion to social movements* (pp. 116–146). London: Blackwell.

Eyerman, R., & Jamison, A. (1989). Environmental knowledge as an organizational weapon: The case of Greenpeace. *Social Science Information, 28*(1), 99–119. doi:10.1177/053901889028001005

Facebook (2013). Retrieved October 2013, from https://newsroom.fb.com/Key-Facts.

Gamson, W. (1992). *Talking politics*. Cambridge: Cambridge University Press.

Gamson, W., & Wolfsfeld, G. (1993). Movements and media as interacting systems. *Annals of the American Academy of Political and Social Science, 528*(1), 114–125. doi:10.1177/0002716293528001009

Garrett, R. K. (2006). Protest in an information society: A review of literature on social movements and new ICTs. *Information, Communication and Society, 9*(2), 202–224. doi:10.1080/13691180600630773

Gillmor, D. (2004). *We the media: Grassroots journalism by the people, for the people*. Sebastopol, CA: O'Reilly.

Hansen, A. (1993). Greenpeace and press coverage of environmental issues. In A. Hansen (Ed.), *The mass media and environmental issues* (pp. 150–178). Leicester: Leicester University Press.

Hansen, A. (2010). *Environment, media and communication*. London: Routledge.

Harlow, S. (2012). Social media and social movements: Facebook and an online Guatemalan justice movement that moved offline. *New Media & Society, 14*(2), 225–243. doi:10.1177/1461444811410408

Harrild, L. (2010, October 27). Lessons from the palm oil showdown. *The Guardian*. Retrived from http://www.theguardian.com/sustainable-business/palm-oil-greenpeace-social-media

Hestres, L. E. (2014). Preaching to the choir: Internet-mediated advocacy, issue public mobilization, and climate change. *New Media & Society, 16*, 323–339. doi:10.1177/1461444813480361

Iyengar, S. (1994). *Is anyone responsible? How television frames political issues*. Chicago, IL: University of Chicago Press.

Kaufman, L. (2011, December 18). Environmentalists get down to earth. *The New York Times*. Retrived from http://www.nytimes.com/2011/12/18/sunday-review/environmentalists-get-down-to-earth.html?_r=0

Kottman, T. (2001). Adlerian play therapy. *International Journal of Play Therapy, 10*(2), 1–12. doi:10.1037/h0089476

Lemert, J. (1977). Journalists and mobilizing information. *Journalism Quarterly, 54*, 721–726. doi:10.1177/107769907705400408

Lester, L., & Hutchins, B. (2009). Power games: Conflict, politics and news. *Media, Culture & Society, 31*, 579–595.

Lim, M. (2012). Clicks, cabs, and coffee houses: Social media and oppositional movements in Egypt, 2004-2011. *Journal of Communication 62*, 231–248. doi:10.1111/j.1460-2466.2012.01628.x

Lipschutz, R. D., & McKendry, C. (2011). Social movements and global civil society. In J. S. Dryzek, R. B. Norgaard, & D. Schlosberg (Eds.), *The Oxford handbook of climate change and society* (pp. 369–383). Oxford: Oxford University Press.

Lorenzoni, I., Nicholson-Cole, S., & Whitmarsh, L. (2007). Barriers perceived to engaging with climate change among the UK public and their policy implications. *Global Environmental Change, 17*, 445–459. doi:10.1016/j.gloenvcha.2007.01.004

McCarthy, J. D. (1996). Constraints and opportunities in adopting, adapting, and inventing. In D. McAdam, J. D. McCarthy, & M. N. Zald (Eds.), *Comparative perspectives on social movements: Political opportunities, mobilizing structures, and cultural framings* (pp. 141–151). New York: Cambridge University Press.

Nerlich, B., Koteyko, N., & Brown, B. (2010). Theory and language of climate change communication. *Wiley Interdisciplinary Reviews: Climate Change, 1*(1), 97–110. doi:10.1002/wcc.2

Netrebo, T. (2012). *Drawing the environment: Construction of environmental challenges by Greenpeace and WWF via Facebook* (MA Thesis). University of Upsala, Upsala.

Newell, P. (2000a). *Climate for change: Non-state actors and the global politics of the greenhouse*. Cambridge: Cambridge University Press. doi:10.1017/CBO9780511529436

Newell, P. (2000b). Environmental NGOs and globalization: The governance of TNCs. In R. Cohen & S. M. Rai (Eds.), *Global social movements* (pp. 117–133). London: Continuum.

Parry, T. (2011, December 16). Facebook's green vow. *The Mirror*, p. 30.

Pattenden, M. (2011, May 13). Does Greenpeace protest too much? Established 40 years ago, Greenpeace is now the world's most powerful NGO. But some now question its anti-nuclear goal. *The Times.* Retrived from http://www.thetimes.co.uk/tto/life/article3017606.ece

Pickerill, J. (2003). *Cyberprotest: Environmental activism online.* Manchester: Manchester University Press.

Ray, A. (2013). The real data on Facebook vs. Google+ (and other social networks). *Socialmedia today.* http://socialmediatoday.com/augieray1/1613711/real-data-facebook-vs-google-and-other-social-networks-interactive-infographic (Retrieved September, 2013).

Roslin, A. (2011, June 4). Www.Energy.Hog.Com. *The Gazette*, p. B1.

Ryan, C. (1991). *Prime time activism: Media strategies for grassroots organizing.* Boston, MA: South End Press.

Schäfer, M. (2012). Online communication on climate change and climate politics: A literature review. *Wiley Interdisciplinary Reviews: Climate Change, 3,* 527–543.

Scott, W. A. (1955). Reliability of content analysis: The case of nominal scale coding. *Public Opinion Quarterly, 19,* 321–325. doi:10.1086/266577

Stoddart, M. C. J., & MacDonald, L. (2011). "Keep it wild, keep it local": Comparing news media and the internet as sites for environmental movement activism for Jumbo Pass, British Columbia. *Canadian Journal of Sociology, 36,* 337–360.

Taylor, V., & Van Dyke, N. (2009). "Get up, stand up": Tactical repertoires of social movements. In D. A. Snow, S. A. Soule, & H. Kriesi (Eds.), *Ecological resistance movements: The Blackwell companion to social movements* (pp. 262–293). Oxford: Blackwell.

Warkentin, C. (2001). *Reshaping world politics: NGOs, the internet, and global civil society.* Lanham, MD: Rowman & Littlefield.

Wexley, K. N., & Nemeroff, W. F. (1975). Effectiveness of positive reinforcement and goal setting as methods of management development. *Journal of Applied Psychology, 60,* 446–450. doi:10.1037/h0076912

Whitmarsh, L., O'Neill, S., & Lorenzoni, I. (Eds.). (2011). *Engaging the public with climate change: Behavior change and communication.* London: Routledge.

Whitmarsh, L., O'Neill, S., & Lorenzoni, I. (2013). Public engagement with climate change: What do we know and where do we go from here? *International Journal of Media & Cultural Politics, 9*(1), 7–25. doi:10.1386/macp.9.1.7_1

Yearley, S. (2008). Environmental action groups and other NGOs as communicators of science. In M. Bucchi & B. Trench (Eds.), *Handbook of public communication of science and technology* (pp. 159–171). London: Routledge.

# Index

Printed and bound by CPI Group (UK) Ltd, Croydon, CR0 4YY

23/10/2024

01778254-0011